Ruud Kleinpaste's
BACKYARD
BATTLEFIELD

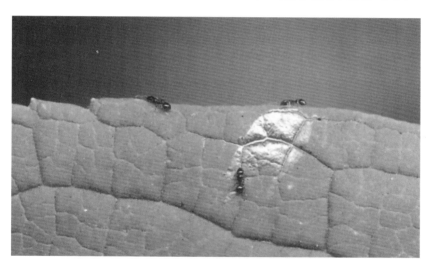

RANDOM HOUSE
NEW ZEALAND

This book is dedicated to my gorgeous and clever wife Julie,
who not only offered useful lectures in English grammar, but also
opened my eyes to the beauty of gardening, the form of plants,
the colours of flowers and the shapes and textures of foliage.
Previously I had seen these things only as insect fodder.

National Library of New Zealand Cataloguing-in-Publication Data
Kleinpaste, Ruud.
Backyard battlefield / Ruud Kleinpaste.
Includes index.
ISBN 1-86941-691-0
1. Plants, Ornamental-Diseases and pests-New Zealand
-Handbooks, manuals, etc. 2. Plants, Ornamental-Diseases and
pests-Control-New Zealand-Handbooks, manuals, etc. 3. Garden
pests-Control-New Zealand-Handbooks, manuals, etc. I. Title.
635.920993-dc 22

A RANDOM HOUSE BOOK
published by
Random House New Zealand
18 Poland Road, Glenfield, Auckland, New Zealand
www.randomhouse.co.nz

First published 2005

ISBN 1 86941 691 0

Design: Katy Yiakmis
Photographs: Ruud Kleinpaste; Powelliphanta photo page 78 by P. Topping
Back cover photograph: Tristan Kleinpaste
Printed in China

Contents

Foreword

I THINK IT IS only fitting to commence this book with a quote from my good old father; he's been dead now for a decade or so but he always had some sublime advice for anyone that would listen.

'Ruud, it doesn't matter what you're going to be doing with your life; as long as you learn how to grow your own food, you'll always have a skill to keep yourself alive on this planet, in any circumstance.'

That really was a bit steep coming from someone who had gangrene fingers and wouldn't know the difference between a carrot seedling and a weed, but he was right, you know (as he was so often). Most of us are into growing domesticated and wild plants, either for fun or for profit, and all of us find it bewildering when disaster strikes, causing the demise or disfigurement of our charges.

Plant health is the topic I seem to have been surrounded by since I started work at MAF in the early 1980s. I have always been fascinated by the hugely complex interactions that take place in our gardens and horticultural ventures: interactions between soil conditions, nutrition, acidity, fungi and bacteria, nematodes, weather, viruses, sun and shade, mites, myriapods and other invertebrates. It truly is a battlefield out there!

There's no way that all these interactions and battles can be described in one book, or even a moderately long series of books. For starters, a lot of the strands of the 'web of life' are still largely

unknown, even in that tiny patch of soil you call your garden. That in itself is a compelling reason for me to avoid the use of aggressive chemicals in my quarter-acre paradise. We still haven't got a clue as to some of the repercussions of our control efforts.

Some really misguided souls still think that economics, industrial progress and politics are 'the real world'. Of course, economics, progress and politics are important. But, as an old American Indian proverb says so clearly:

Only after the last tree has been cut down,

only after the last river has been poisoned,

only after the last fish has been caught,

only then will you find that money cannot be eaten.

In this book you will find glimpses of the battle that takes place in your backyard. It deals mostly with entomological subjects (the insects), just one of the groups of players in the plant health game.

It's the group I have come to most love and respect. These small critters literally are the salt of the earth. They regulate soil fertility and are very complex entities in the myriad food chains that live outside your kitchen door.

You don't need to switch on the television and watch *Our World* to see what they do for a living; a torch and a pair of eyes is often all you'll need to see it live in your own backyard.

And my good old dad has the last word yet again: 'Son, it's the only backyard we've got . . .'

Introduction

THE FASCINATING THING ABOUT Nature is that it is an extremely complex system which can be viewed from a number of different angles, ecological as well as historical. These angles can be superimposed on human acts of interference, be it horticulture, monoculture or permaculture. All these 'cultures' are connected to our civilisation — even the yeast culture that grows in between the digits of our hind limbs.

Be it gardening or growing crops on small acreages, working with stock or other suitably domesticated animals, it definitely pays to keep a few fingers on Nature's pulse.

And, to be quite honest, that's the only way to go in this world. If you can work out — or better still, predict — Nature's cycles, you'll be able to work with the tides . . . the ebbs and the flows.

Life cycles, generations, over-wintering cycles, aestivation cycles, oviposition cycles and even outbreak cycles — they're all there in the textbooks. Some of the cycles are so beautifully timed that you can almost set your watch by them; well, perhaps not your watch but certainly your thermometer.

Most invertebrates will spend the winter in diapause, or at the pupa or egg stage. This makes sense as cold-blooded critters simply rely on the external temperatures for all their living and loving. So when it's cold, there ain't much activity . . .

With spring come the first insects, and as we all know, with the first insects come

the first bits of feeding damage on our plants. The more spring progresses into summer, the more beasts will be aiming for your crops, and the more control strategies will be needed to combat them. This book aims to help you identify the guilty parties and to deal with them in the most eco-friendly way possible so that you can win the backyard battles.

Chewers, suckers and raspers

ENTOMOLOGISTS CAN OFTEN IDENTIFY within a few seconds, what beastie or what clan of critters has been responsible for the mayhem caused on your plants, simply by looking at the tell-tale signs. If you gather enough diagnostic evidence you'll find out what the causal organism is likely to be. Once you know what the organism is, you know its life cycle and, hence, the best method of prevention or control.

When we look at damage patterns we can concentrate on the three main groups of vandals: the chewers, the suckers and the raspers. It may come as no surprise that these groupings relate to the shape, size and mode of action of their members' mandibles.

Chewers

Let's start off with the chewers — the group that most people will find easy to identify. Chewing pests hail from a limited range of insect Orders: the main groups are beetles and weevils, caterpillars, weta and katydids, plant-feeding wasps, and some maggots. In the non-insect department we'll often encounter slugs and snails as chewing pests.

Chewing invertebrates are endowed with opposable mandibles; a bit like ours. They can bite bits off leaves and other plant parts, which they masticate and swallow before digesting them and releasing the waste as frass — a nice entomological term for excrement.

There's a whole range of symptoms that a gardener will have no trouble

identifying: first of all, the chewer removes bits of plant material, which results in holes; holes in flower petals, holes in leaves, on the edges of leaves (scalloping or notching), or something that can be described as 'shot-holes' in leaves.

Leaf damage often looks pretty spectacular, and it is the first thing that gardeners and growers notice. But is it really as important to the plant's health as it appears?

Chewing damage is — if you think about it — nothing more than a natural form of mechanical pruning, and most well-established plants can easily withstand a bit of that.

In most circumstances the natural pruning performed by leaf-chewing invertebrates is really of no concern to the overall health of a plant.

Whether or not it looks attractive is a matter of personal opinion!

However, some chewers are quite tricky customers, especially when they irritate their host to such an extent that the poor plant throws up a whole lot of quickly-grown plant cells in an attempt to encapsulate or isolate the pest. These growths are called galls. An enormous range of plant species can produce galls and, I hasten to add, not only because of chewing invertebrates. Some galls are caused by sucking creatures, others by rasping villains, and even diseases can cause a whole heap of galls as symptoms.

However, if you open a gall and a small larva wriggles around inside, you can bet your boots that it's one of the chewers at work.

Suckers

When we look at the impact of invertebrates on the health of their hosts, chewing damage is merely chicken feed when compared to the work of the suckers, which can debilitate the healthiest of plants. To get an impression of the modus operandi of sucking creatures we need to look at how a vascular plant operates.

Water and nutrients are taken up by the plant's roots and transported through the wood to the leaves. Leaves take up carbon dioxide and they utilise the sun's rays to provide energy for photosynthesis. This process combines the raw nutrients with the carbon dioxide and produces oxygen and carbohydrates (sugars), which form the building blocks of new plant growth. The sugars are transported throughout the plant in vascular bundles called phloem. And that's quite handy if you are a sucker. All you need to do is use your sharp, tubular mouthparts to hit one of those rich phloem veins . . . then sit back and enjoy the sweet, viscous liquid that will literally flood into your mouth.

Early detection, identification and control of sucking invertebrates is of importance. The first sign to look for is the deposit of honeydew all over a plant (pictured overleaf top). If you find honeydew splattered all over the leaves, just look straight up from there and you'll be staring into the bum of a sucker! The formation of sooty mould, on leaves, twigs and branches (pictured overleaf lower), is another easy-to-spot result of the activity of sucking insects. This

fungus specialises in colonising energy-rich honeydew deposits.

When there are large numbers of 'mainlining' insects, the results can be devastating for a plant. The interception of large quantities of carbohydrates means that the plant won't get its regular share of nutrients. This will translate immediately into a general pattern of ill thrift; deficiency symptoms such as yellowing, purpling, stunting and distortion can result. In heavy infestations, wilting indicates that the liquid nutrients aren't getting through to the growing tips. But one of the most debilitating results of sucking damage is often a virus infection transmitted by the sucking bug itself. Think hypodermic needles and veins and you'll understand the risks plants run with suckers.

Raspers

Raspers also damage their host plants with tubular mouthparts, but in this case the invertebrates are so small that their mandibles do not reach the vascular bundles at all. Instead, they scrape away at cellular level and suck out the contents of individual plant cells. Thrips and mites are the prime examples of raspers.

It is important to accurately identify the damage patterns, as thrips require different chemicals for control from mites: insecticides and miticides respectively.

Some raspers break into special cells within the plant, causing an unexpected range of symptoms. For example, bottle-brush thrips damage the meristem cells of their host, Callistemon. This results in the

---

HONEYDEW

HONEYDEW IS VISCOUS, CLEAR AND STICKY. IT IS THE EXCESS SWEET LIQUID THAT A SUCKER HAS EXTRACTED FROM THE PHLOEM, MINUS ITS NITROGEN AND PROTEINS. HONEY-DEW IS EXCRETED IN A RANDOM FASHION AND IS OFTEN SPLATTERED ALL OVER THE LEAVES AND STEMS OF THE HOST AND, SOMETIMES, ALSO OVER THE PLANTS THAT GROW UNDERNEATH THE HOST. LICK IT UP — IT TASTES WONDERFUL — BUT DON'T THINK ABOUT THE FACT THAT IT HAS GONE THROUGH THE INTESTINAL TRACT OF AN INSECT. AS A PRODUCT OF PHOTOSYNTHESIS, CREATING OXYGEN AND PLANT GROWTH, IT IS THE SOURCE OF ALL LIFE ON EARTH.

new growth being totally distorted and twisted. Certain mite species accomplish the same tricks: broad mite and cyclamen mite are known villains that sabotage the dividing cells in the growing tips of plants.

Predators and parasites

GARDENERS AND GROWERS ARE a switched-on lot, these days. They know that their little paradise is an ecosystem full of plant and animal species, each vying for their own continued survival. While plants do get damaged by invertebrates most, if not all, of these plant pests have natural enemies too, and in a nicely balanced world these enemies tend to keep the plant pests under control; at least to some extent.

When I'm giving talks on 'How to murder your plants more slowly', one of the most commonly asked questions pertains to the identification of the 'goodies' in the garden. In other words, which predators and parasites should we encourage and — more importantly — how do we identify these critters? What do they look like, what do they do for a living, and how can we attract them?

Predators

Predators are invertebrates that actively 'hunt' for prey, capture it in some way, and devour it on the spot. In the greater scheme of things, predators need to eat meat so they can stay alive or to successfully complete their larval growth stages in order to reach adulthood. In general they feed on live meat — you can't get it any fresher than that. Some predator species are very choosy in their prey (prey-specific predators), while others don't really care what they eat, as long as it's about the right size and shape (generalist predators).

Parasites

True parasites rarely kill their host; they simply feed on body parts that may not be vital to that host. In fact, parasites make a living out of consuming the nutritional reserves of their host. Good examples of such parasites in the class Insecta are our headlice, bedbugs and mosquitoes. Hair follicle mites, tapeworms and round-worms are non-insect parasites.

The beasts that growers and gardeners tend to revere as 'parasites' are better termed parasitoids, as these useful critters ultimately kill their hosts. As such they have more in common with predators than with parasites per se. Our parasitoids are mainly wasps and flies. The adults locate prey for their larval offspring, while feeding on nectar and pollen themselves.

In most instances the relationship between insect host and parasitoid is a unique one. Host-specific relationships hinge around a perfect synchronisation of life cycles: there's no point for a female parasitic wasp to be active (and searching)

when the required host caterpillar happens to be on 'time-out' in the egg stage. There are also some parasitoids that don't care what their young are feeding on, and therefore have a wide range of possible targets.

Parasitoids also go through a life cycle, in which the feeding and growing larval stages live inside the body of the host, where they have to be very careful not to kill their living food source on day one. Like true parasites, parasitoid larvae initially feed on the non-essential tissues inside the host, which keeps the landlord alive, but not alive and well: often an infected host shows signs of distress or a marked decrease in activity and development. Only at the final stages of development will the parasitoid kill its host, when it triumphantly exits the lifeless body, to pupate.

Despite everything, predators and parasites are the really good guys in the backyard battle. They are known as beneficials, and they fascinate me. The good news is that these invertebrates are usually able to sustain themselves unaided and carry out their biological control almost unnoticed in your botanical patch. And more importantly: they do it all for free!

I think it is fair to say that each and every living creature on this planet has its own suite of natural enemies, and our pest insects are no exception. Responsible and

COMPANION PLANTING

THE BEST PLANTS FOR PARASITOIDS SUCH AS HOVERFLIES AND LACEWINGS ARE *PHACELIA TANACETIFOLIA* (BLUE TANSY, PICTURED RIGHT), AND MEMBERS OF THE UMBELLI-FEROUS FAMILY APIACEAE SUCH AS QUEEN ANNE'S LACE, WILD CARROT, FLOWERING PARSLEY, EVEN HEMLOCK. BUCKWHEAT IS A GREAT ATTRACTANT FOR PARASITOIDS. MOST FLOWERS WILL PROVIDE SOME POLLEN AND NECTAR FOR YOUR BENEFICIALS, SO MIX AND MATCH VEGGIES AND FLOWERS, OR FRUIT CROPS AND FLOWERS. THEN ALL YOU HAVE TO DO IS SIT BACK, RELAX AND LET THE GOOD GUYS DO MOST OF THE WORK.

sustainable growers and gardeners know exactly what I am talking about here. The popularity of natural enemies is on the increase and biological control has become more than a buzzword (excuse the pun).

We've come a long way with understanding our good guys. If you are a gardener or lifestyle blocker who wants to grow a few crops with a minimum of sprays or bother, the good guys will come to your place with a wee bit of encouragement and here is how:

1) cut out the artificial sprays;
2) sow flowering plants that will give your good guys plenty of pollen and nectar so they can reproduce and lay many eggs in and amongst your pests; and
3) maintain 'biodiversity' by mixing up as many crops and flowers as possible.

Control strategies

CONTROL STRATEGIES — SOUNDS A bit like sophisticated warfare. Indeed it can be, and preparation in spring could be the key to putting your defence mechanisms in place. There are a number of control strategies recognised on our crop-growing planet.

BIOLOGICAL CONTROL is one of those magical solutions that could solve all our problems in one easy little package — if only it were that simple!

This method uses the 'three Ps': predators, parasites and pathogens, and can be implemented in three different ways:

1) 'Classical biological control' involves introducing natural enemies of a certain exotic pest. Once these enemies are established, they'll do their job restricting numbers of the original pest. A good example of this is the smorgasbord of parasites introduced to New Zealand to help with the control of the dreaded white butterfly on brassicas.

2) 'Augmentation' is the technique whereby we release a whole lot of natural enemies into an area where they are absent (or in low numbers), to help combat a pest insect. The whitefly parasite and predatory mite are prime examples of this augmentation strategy.

3) 'Conservation' means looking after our goodies, by providing them with the right habitats and foods, so that their populations can be maintained in the garden or orchard. This, of course, includes giving the parasites and predators shelter sites and over-wintering places, as well as pollen and nectar sources for the adults. It also means that chemical and organic sprays may need to be restricted or banned.

CULTURAL CONTROL aims at changing the environment, so that pest invertebrates are disadvantaged.

Hygiene is important in this regard because a lot of insects over-winter in crop debris (mites, thrips, some aphid species, passionvine hopper etc). The removal of old crop debris, including mummified fruit, is important. Clearing away alternative host plants and weeds, which facilitate the continued life cycle of pests and diseases, is another key to success.

Crop rotation is useful when pests rely on the same host year after year. By changing crops regularly, the population density of pests in the soil will not reach major proportions.

Intercropping is a great idea to reduce pests, especially when two or three crops work as 'companion plants'. The idea is to make it harder for pests to find their favourite host, simply by 'masking' these hosts in amongst a whole lot of other plants. Good examples are onions inter-

BIOLOGICAL CONTROL WITH INSECT PATHOGENS

BACILLUS THURINGIENSIS VAR. *ISRAELENSIS*, ALSO KNOWN AS BTI (MARKETED AS VECTOBAC WDG AND AVAILABLE IN 500 GRAM QUANTITIES), IS A GREAT WEAPON AGAINST MOSQUITO LARVAE IN WATER. IT KILLS THE WRIGGLERS WITH A MASSIVE BOUT OF DIARRHOEA.

BACILLUS THURINGIENSIS VAR. *KURSTAKI*, ALSO KNOWN AS BTK OR THURICIDE (MARKETED AS DIPEL, AGREE, OR FORAY), CONTAINS A NATURAL BACTERIAL GUT DISEASE OF CATERPILLARS. IT ACTS WITHIN 3 TO 4 HOURS TO STOP THE CATERPILLAR'S APPETITE. IT WILL DIE A FEW DAYS LATER OF DEHYDRATION AND DIARRHOEA. THE GREAT THING ABOUT BTK IS THAT IT WILL NOT AFFECT NON-TARGET SPECIES SUCH AS PREDATORS AND PARASITES. HUMANS ARE ALSO OUT OF THE DANGER ZONE, UNLESS YOUR GUT RESEMBLES THAT OF A CATERPILLAR.

SPINOSAD WAS DISCOVERED IN THE 1980s IN THE SOIL SAMPLES OF AN ABANDONED CARIBBEAN RUM-STILL. A BACTERIUM, IDENTIFIED AS *SACCHARAPOLYSPORA SPINOSA*, METABOLISES AND EXCRETES A BIOLOGICALLY ACTIVE SUBSTANCE THAT CONTROLS BUTTERFLIES AND MOTHS, FLIES AND THRIPS. MARKETED UNDER THE NAME SUCCESS, IT IS A REMARKABLE NATURAL INSECTICIDE. GOOD NEWS INDEED, AS IT WORKS WELL ON LOOPERS, WHITE BUTTERFLY, TOMATO FRUITWORM AND LEAFROLLERS!

planted with carrots (carrot rust fly hates onions while luckily, onion fly is not impressed by the smell of carrots) and the combination of tomatoes and cabbages (which tends to cause a reduction in pest levels of white butterfly caterpillar).

And then there are the reports of basil working against thrips on tomatoes, and garlic against just about anything (weevils, aphids and mites).

Companion planting can take the form of sacrificial planting. For example, the tomato fruitworm (*Helicoverpa armigera*) is also known as the corn ear worm. Caterpillars will tunnel into both kinds of 'fruit'. But when you plant corn in between the tomato rows, the caterpillars are recorded as preferring to go for the developing corn ears, leaving the young tomatoes alone. It must be stressed, however, that a lot of the reputed companion plant combinations have not been scientifically proven, and may be wishful thinking on the part of the gardener.

What has been proven to work well is the interplanting of crops with flowering plants attractive to the adult predators and parasites. The flowers contain a wealth of pollen and nectar, useful for the wasps and flies to develop their reproductive potential. The more eggs they can lay inside and on our pests, the better percentage of biological control we'll get.

Timing is also of importance. The date you sow a crop could be altered to escape the emergence or flight period of a particular pest. A good example is the carrot aphid, which flies in mid- to late spring; you either make sure your carrots are harvested by then, or you sow early in December.

Some forms of cultural control rely on knowledge of the pest's life cycle. Cultivation of the soil after cropping destroys a good number of larvae and pupae of pest insects, and exposes them to birds and other predators.

Altering the environment to deter pests can be as simple as raising the relative humidity in the evening: for example, two-spotted spider mites love warm, dry conditions, so regularly mist the underside of leaves in the evening with water and you'll give them cold, wet feet, which they hate, and consequently populations will crash. But when you get into this kind of stuff, it may pay to get some preventative fungus

spray on first, as the foliage will remain wet for prolonged periods of time, increasing the chances of fungal and bacterial infection.

Inside glasshouses, temperature, ventilation, relative humidity and shade can be controlled to the plant's advantage, and the pest's disadvantage. Even the composition of potting or growing mixes can be altered to reduce the debilitating activities of soil-dwelling pests.

The use of appropriate mulches can have a negative influence on pests. Reflective mulches (rolls of aluminium foil in between the rows) tend to reduce aphid and thrips numbers in some crops, presumably because short-wavelength light is reflected upwards, thereby obscuring or disguising the palatable plants. Non-reflective mulches (coloured hessian or shade cloth) around plants appear to minimise the colour contrast between plants and surrounding terrain, also making them more difficult to detect. Of course, these mulches will not deter any pests that find their host plants via smell or other chemical cues!

The most dramatic alteration of the 'food environment' is a simple genetic manipulation. Resistant or 'tolerant' crops have been around for evolutionary ages, and crop scientists have researched this phenomenon for decades. Some of these specially bred varieties are available to the gardener; look out for them because they do have an effect.

PHYSICAL CONTROL means that you physically destroy the pests or deny them access to your prized plants.

Digital control employs the two most nimble digits of your hand: pick the blighters off and squash them between finger and thumb, or feed them to the chooks. Squashing on a large scale can mean the trampling hooves of a sizeable

BUGS FOR SALE

THE NEXT LOGICAL STEP IN 'NATURAL' PEST CONTROL IS THE EMERGENCE OF SMALL COMPANIES THAT ACTUALLY BREED BENEFICIALS FOR SALE. IT'S ALREADY HAPPENING IN NEW ZEALAND.

TWO COMPANIES ARE INVOLVED IN THIS VENTURE: BIOFORCE LTD, BASED IN KARAKA, SOUTH AUCKLAND, AND ZONDA RESOURCES LTD IN HASTINGS AND PUKEKOHE. BOTH STRESS THAT THEIR MAIN TARGET CLIENTS ARE COMMERCIAL GROWERS, BECAUSE OF THE FACT THAT INTEGRATED PEST MANAGEMENT (IPM) AND THE USE OF BENEFICIALS IS A RATHER COMPLEX AFFAIR, REQUIRING BOTH KNOWLEDGE AND SKILLS TO MAKE IT SUCCESSFUL.

herd of stock on grass grub-infested pasture after a decent autumn rain (mob-stocking), or a heavy, water-filled roller on a lawn.

Pruning branches with stemborer inside, or stems with scale insects, is as simple as the removal of over-wintering oviposition sites of passionvine hoppers, as long as you know what you are looking for and time the surgical procedure in accordance with the pest's life cycle.

Pheromone traps and pheromone traps lure the male moths of a particular species away from the waiting females, which can result in a significant reduction in fertilised eggs.

Coloured sticky traps immobilise the flying pests (whitefly on yellow traps, thrips — pictured below left— on blue sticky traps) on the tacky surface, which apparently takes away any desire and ability to reproduce.

Physical barriers can also be contemplated. While we are on a sticky subject, some people report great success with sticky bands around stems and trunks, to stop crawling insects getting at the goodies higher up. Fine mesh screens over doors and windows of glasshouses restrict the entry of whitefly and other flying pests, and lightweight crop covers achieve similar results for field-grown crops. An added bonus is that these covers can also protect certain crops from light frosts.

Some plant pests fly at low altitude, simply because they lay their eggs at or near soil level (bulb flies and onion flies at the neck of bulbs and onions, carrot fly on the exposed top of carrots, etc). A simple one-metre tall 'fence', made from shade cloth, is surprisingly effective in keeping these kinds of flying pests out of the vegetable plot. Somehow they don't seem to cotton on to the idea of flying over the fence, and hence they miss their target altogether. A similar barrier can be made from tall, dense plants around the veggie garden — a little bit like an impenetrable hedge. And wouldn't it be nice if these plants were bearing blue flowers full of pollen and nectar for the beneficials?

ORGANIC SPRAYS use compounds that are derived from plant extracts or other natural ingredients. This does not mean that these sprays are '100% safe' for humans. Each one of them contains chemicals with their own natural toxicity to mammals.

The problem with some organic sprays is that they are not very selective; they will kill a whole range of invertebrates: the pests and the beneficials! This means that we still need to know exactly what we are doing, and time and aim our sprays to perfection, to minimise the damage to non-target species.

Some of the better known botanical insecticides are pyrethrum, Derris dust and neem oil. The first two are powerful insecticides with a rather quick action and not much discrimination as to who gets killed. Only neem oil seems to be friendly to beneficial predators and parasites.

Most botanical sprays break down rapidly in the environment. The advantage of this is the lack of residue on crops. The disadvantage is the stuff will not work for prolonged periods, and repeat sprays are frequently required. Moreover, botanical insecticides tend to be ineffectual on soil insects.

Spraying oils come in all sorts of shapes and sizes. The most commonly used is a refined mineral oil. This oil works simply by smothering the pest (and other) insects, and clogging up their respiratory system. On mealybugs (pictured below) it breaks down the protective white waxy barrier of meal on the insect's body, and with scale insects that live under their own excreted, waterproof 'limpet' the oil creeps under the protective cover and catches the inhabitant unawares. Repeat sprays are often necessary.

Mineral oil comes in 'winter' and 'summer' forms, indicating that the concentrations need to be altered according to the season. Indeed, mineral oils can damage certain plants under high sunlight intensities.

Insecticidal soaps are another fabulous organic material. These contain the potassium salts of fatty acids, which make them rather abrasive to the skins of soft-bodied invertebrates such as aphids, juvenile thrips, whiteflies and scale crawlers. They leave no active residue and are generally not harmful to larger, hard-bodied invertebrates. They can, however, be toxic to some plants, such as ones with hairy leaves, especially when applied in bright sunlight.

Lime sulphur is the favourite winter clean-up spray of many rose growers. It will kill a wide range of over-wintering pests, such as mites, aphids, mealybugs and scale insects, but it also has a knack of damaging some plant tissue and even lichens. This is why it tends to be used in winter when many plants are dormant.

Slugs and snails can be baited with pellets laced with iron chelates, or haphazardly lured with stale beer in containers sunken into the ground. Here they die a relaxed death in a drunken stupor.

Then there are all the other organic mixtures that have been doing the rounds for the past hundreds of years. They involve homemade concoctions of rhubarb, tomato, garlic and onion, and go as far as eucalyptus leaves, and ground, dried pyrethrum flowers — some even use cayenne pepper! To be quite honest, I haven't tried them and I haven't seen any fair-dinkum efficacy data that convinces me of their usefulness. All I can say is: 'Give it a go, but be careful not to poison yourself!'

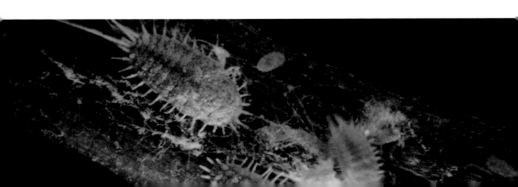

The fear of poisoning one's self, plus the garden, the environment, and the universe, is usually associated with CHEMICAL CONTROL. There is a bewildering array of synthetic insecticides available, but their popularity seems to be waning somewhat.

If you must use them, there are some excellent books and advisory leaflets about the chemical control of invertebrate pests. Make sure you also consult the various label claims and carefully follow the methods of application and directions of use printed on each container. Chemical control can be a tricky business. The current New Zealand standard for agrichemical users is NZS 8409 (Management of Agrichemicals). Some materials have a damaging effect on plants. When selecting a chemical control compound it pays to get an informed identification of the pest first — decision-making will then be a lot easier.

Some chemical compounds can be used as the active ingredient for baits, where the idea is to target only the pests, and minimise kill of non-target species.

Systemic pesticides are translocated through the plant. Some materials only go from the top surface of the leaf to the underside, whereas others are taken up by the sap stream and travel through the plant in this manner. It comes as no surprise that systemics are usually applied to combat chewing, rasping and sap-sucking invertebrates living inside the plant, protected in nooks and crannies, or hidden from view. Unfortunately, systemics often tend to be the first compounds to which invertebrates develop resistance. Use them wisely — regular rotation of active ingredients, preferably from a different chemical group, helps delay the development of resistance or tolerance.

Contact insecticides need to be applied directly onto the pest for the active ingredient to work effectively, or else deposited as a residue that the pests come into contact with at a later stage. Good coverage is often essential.

Fumigant insecticides are delivered as a fine spray, but reach most nooks and crannies as a result of their gas-like action.

Most insecticides act as a stomach poison, and interfere with the nervous system of the invertebrate.

It is important to realise that a treatment with chemical (or even some organic) insecticides can destroy all your good biological control activities in one ill-timed or ill-directed spray.

In my view, there is a place for the responsible use of chemicals to control insect pests in this world, but whether or not that place is the home garden is a matter for vigorous debate and Yates is slowly replacing old-fashioned heavy compounds with 'softer' chemicals. In any case, there are plenty of alternative techniques and ideas at your disposal to discourage or control most, if not all, of the pests that can make your life a misery.

INTEGRATED PEST MANAGEMENT combines a number of control strategies that work together to obtain the best possible control, consistent with economic, ecological and toxicological requirements. In gardens and small plots this may well translate into utilising natural enemies as best we can, and adding cultural and physical control techniques (perhaps even organic and chemical compounds) to further reduce the damage caused by plant pests.

The first question, though, is: what are our plant-damage thresholds? We cannot escape the fact that the species *Homo sapiens* must live within its own ecosystem, even when it comes to growing food and ornamental crops. Some plant damage can be expected and must be tolerated. I've always wanted to redefine the concept of the 'perfect plant' as one which has quite a bit of wildlife on it . . . and, of course, it grows in a healthy garden! For beneficial organisms to thrive, you must have pests. Sorry to be a spoilsport, but that's ecological rule number one.

For successful IPM you'll need to know a bit about the level of pests on each crop in your area. You'll need to find out about their life cycles and timing, their flight habits and mode of dispersal, their likes and dislikes (alternative hosts, or plants that repel them), and you'll need to spend some time monitoring your crops.

Whatever you do in your garden, there are myriad options for pest prevention and pest control, and most of these can be incorporated when planning in spring for a healthy garden.

Spiders and mites

THE ARACHNIDS ARE A fascinating group of intvertebrates, made up of an assembly of different and perhaps unexpected characters. Mites, harvestmen, whip scorpions and ticks belong to this class, as well as, of course, spiders.

Spiders differ from insects in that they have eight legs — as opposed to six — and only two body segments. Their head and thorax are fused into one segment, called the cephalothorax, which literally means 'head-thorax'. The abdomen contains the digestive system, genitalia, and the silk glands and spinnerets.

In New Zealand we have about 2500 different species of spiders, with only half of them officially named. And all these species have their own preferred prey and their own unique way of catching it. If you take a good look in your garden, you're sure to find a spider or two or, at least, evidence of their deadly snares.

Life cycle

Spiders go through a whole range of rituals when boy meets girl. Usually the male is smaller than the female, which means he has to be careful not to be mistaken for prey. After a prolonged period of spider-species-specific foreplay, the actual act of insemination is quite tricky — it's a process that can be best described as 'indirect fertilisation'.

Males possess modified palps, charac-terised by a swollen segment that is

hollow and has a kind of spring mechanism inside. These swollen palps are used to distinguish male spiders from females. The male palps are pre-charged with sperm, which means that the hollow organs work something like a pipette. The male inserts one — or both — of these palps into the female's genital opening and runs like hell.

Depending on the species, the female will spin a silken cradle or sac for her eggs and guard them until they hatch. Spiderlings that hatch usually stay in the parental egg sac for the first few moults, before biting through the silk and escaping into the world. Some species then commence their ballooning flights in order to disperse.

Ballooning is another silk-driven miracle — young spiders climb to the top of vegetation and let a thin strand of silk escape from their spinnerets. This silk is then picked up by the breeze which, when strong enough, will carry the spiderling to great heights and for many miles.

Each autumn, paddocks full of spider-gossamer can be seen around the country — millions and millions of spiderlings are ready for their first ballooning flight in the dawn of a dewy morning. It is a spectacular sight if you are lucky enough to be in the right place at the right time.

Air travel by spiders is a well-known phenomenon. If you could stick a butterfly net out of the window of a plane, you'd catch ballooning spiders at altitudes of 1500 and even 3000 metres. Ships in the middle of the ocean have reported spiderlings appearing from seemingly nowhere, and each year we get a number of illegal arachno-immigrants from Australia arriving under their own steam.

Benefits

New Zealand is one of the few civilised countries in the world where you can still find products specifically marketed for killing spiders. Isn't it time we woke up?

Cicadas, wasps, beetles, moths, caterpillars and flies: just about every garden pest is on the menu of some spider or other. Arachnids are the true goodies of the garden.

SILK

SPIDER SILK IS A REMARKABLE SUBSTANCE. MOST SPIDERS PRODUCE SILK IN ONE FORM OR ANOTHER. IT COULD BE A SIMPLE DRAG-LINE (USEFUL AS A BUNGY CORD WHEN A TRICKY DESCENT IS REQUIRED), OR A HANDY-WRAP FOR EGGS. AND THEN THERE ARE TRIP-LINES, TRAPDOORS, TUNNEL-LIKE RETREATS, AND THE UBIQUITOUS FULL-BLOWN WEBS. SILK IS A PROTEIN, PRODUCED AS A VISCOUS FLUID BY VERY EFFICIENT SILK GLANDS INSIDE THE SPIDER'S ABDOMEN. THESE GLANDS HAVE DUCTS THAT CONNECT THE GLANDS TO VARIOUS SMALL NIPPLES (OR SPIGOTS) SITUATED ON THE SPINNERETS, WHICH ARE THOSE SMALL KNOBS, USUALLY SIX IN TOTAL, TOWARDS THE APEX OF THE ABDOMEN. SCIENTISTS HAVE IDENTIFIED NINE DIFFERENT KINDS OF SILK, EACH WITH ITS OWN SPECIFIC PURPOSE.

Australian redback

THE AUSTRALIAN REDBACK (LATRODECTUS *hasselti*; pictured below) has been found in New Zealand — here and there — for at least twenty years, and it looks as if it may indeed be established. This spider loves to make its home in the vicinity of human dwellings, a habitat which is best described as a peri-domestic environment: in messy sheds, under houses, in stacks of fire wood, under bridges . . . I'm sure you get the picture.

We know the redback is most famous across the Tasman for hiding under the rim of the toilet seat in the archetypal outdoor dunny. The fact that this remarkable adaptation is a brilliant example of opportunism on the part of the spider is often lost on the slightly nervous outdoor dunny user. These dunnies have a habit of attracting flies, and flies appear to be among the favourites on the menu of redbacks. So if you are able to construct a decent web under the rim of the toilet seat, you can not only eat like a king, but also practice a form of common-sense niche marketing.

A word of warning: if you are male, it is advisable not to sit down on an outdoor dunny. Australian statistics show that Madam Redback will have a go at anything that appears below the rim of that seat!

Just in case you're wondering: you can always tell a Latrodectus species because it has a bright orange-red hourglass mark on the underside of its abdomen.

Australian two-spined spider

THERE ARE SPIDERS THAT recycle the protein of their silk by eating it. A good example is the Australian two-spined spider (*Poecilopachys australasia*). This beast belongs to the orbweb spider Family Araneidae and arrived in New

Zealand, quite by accident, some 30 years ago. It is a real stunner and everybody who finds a female of this species in their garden believes that an alien has invaded their quarter-acre paradise.

The colours are a vibrant yellow, dark red and green, and the carapace or abdomen is punctuated by two distinct spines. The puny male of the species looks boring and spineless, and is about one-fifth the female's size.

Poecilopachys loves to live in citrus bushes, camellias, or anything else with sizeable, shiny leaves. During the day it sits motionless on the underside of a leaf, and despite its colours it can be pretty hard to find. At night the female spins a large, rather incomplete orbweb that acts as an aerial filter for flying insects. The preferred food is small moths, especially leafrollers.

When the morning sun peeps over the horizon, the two-spined spider literally removes all evidence of her nocturnal web by hauling it in and devouring the silk, complete with some of the catch. The next evening the whole process starts again.

The egg sac of this species is quite distinct. It is a spindle-shaped container made from sturdy, mid-brown silk, suspended by fine strands between some leaves. To hatch, the spiderlings chew a neat, round hole in the outer skin of the egg sac.

ARACHNOPHOBIA!

THE AVONDALE SPIDER IS THOUGHT TO HAVE ARRIVED ABOUT ONE HUNDRED YEARS AGO, WITHIN THE STACKS OF AUSTRALIAN HARDWOOD TIMBER USED TO CONSTRUCT THE AVONDALE RACECOURSE BUILDING. OUT IN THE SUBURBS IT LIVES IN SHEDS, UNDER HOUSES, IN WALL CAVITIES, OR UNDER PEELING STRIPS OF WATTLE BARK IN GARDENS AND RESERVES. THIS SPIDER HAS THE GRAND ABILITY TO MAKE ITSELF REALLY FLAT, SO IT CAN SQUEEZE INTO THE NARROWEST CRACKS AND CAVITIES.

AVONDALE SPIDERS WERE THE STARS OF THE STEPHEN SPIELBERG FILM ARACHNOPHOBIA. THE IRONY IS THAT SPIELBERG COULD NOT HAVE CHOSEN A MORE DOCILE SPIDER SPECIES FOR HIS MULTI-MILLION-DOLLAR PRODUCTION — HIS EXPENSIVE HUMANOID ACTORS WERE NEVER AT RISK OF A SPIDER BITE!

Avondale spider

THERE ARE QUITE A few Australian spider species that have made it to New Zealand on a permanent basis. One of the bigger ones is our mate the Avondale spider (*Delena cancerides*; pictured right). Known in its home country as the red-brown huntsman, this impressive arachnid is generally tolerated there as a beneficial critter, emerging at night from behind the painting on the wall to feed on moths, flies, and roaches. Here in New Zealand the Avondale spider still causes the odd spine-tingling stir when it appears from nowhere in the peri-domestic environments of Avondale, New Lynn and Blockhouse Bay.

The reason Avondale spiders have not spread very far can be explained by their unique, gregarious behaviour — Mum and Dad and all the kids remain together in their ancestral hide-out and hunt and feed as a family. Parents tolerate their offspring at home, so there is no need to disperse in a hurry.

When Delena hunts, it does not use a web or snare — a drag line is all that is needed to find the way back home, and to act as a safety wire for near-misses. This spider is hairy and a large number of special spiny hairs on its legs and body detect minute changes in air currents caused by the movement of potential prey. Usually Delena sits quietly and waits for prey to come close enough, then it pounces — it's a truly spectacular display of lightning-fast reflexes and huge masticating mandibles.

Bird-dropping spider

A RATHER STRANGE-LOOKING relative of the two-spined spider is the bird-dropping spider (*Celaenia* species; pictured below). It looks exactly like its common name, especially when it sits on a patch of silk with one or more of its egg sacs. More often than not you can find these amorphous 'blobs' dangling from a short length of silk, under a leaf.

But the really remarkable thing about this 'orbweb' spider is the fact that it does not make a web at all. The spider literally hangs from a strand of silk at night and captures moths that fly past, by grabbing them with its spiny front legs. The technique sounds simple enough, but what are the realistic chances of a suitably sized moth flying by within 'grabbing distance'?

A clue to this mystery was found by identifying Celaenia's prey — it consisted of male moths, mainly of the leafroller type. This led to the assumption that the spider releases an attractive chemical, similar to the leafroller's sex pheromone — a stunt also performed

by related spiders in Australia. A male leafroller moth 'on final approach' will suddenly find himself grabbed by a predator instead of having the pleasures he desires.

Golden orbweb spider

PROBABLY THE MOST SPECTACULAR species we get from across the Tasman is the golden orbweb spider, (*Nephila edulis;* pictured below). In spring, small spiders balloon to New Zealand, probably driven by areas of low barometric pressure that pick up the travellers in Queensland and drop them off as the fronts move across the west coast of Aotearoa. There they find a place to spin their majestic webs made from strong, golden silk. As the spiders grow their web becomes larger and larger — often reaching the size of a small volleyball net.

These webs, and their large inhabitants — body length 25 millimetres, leg span 120 millimetres — go for the larger brand of insects: cicadas, wasps, and large beetles. Occasionally they feed on small birds. Yes,

the web is strong enough to catch birds.

A few females of *Nephila edulis* are found each summer and autumn in various, widely spaced localities, but so far there is no evidence of it becoming established in New Zealand — the tiny males have never been found here.

Grey house spider

THE GREY HOUSE SPIDER (*Badumna longinqua;* pictured below) is probably the most common spider in New Zealand, but it appears to have also come from Australia. It is dark grey to black, with light grey hairs on its abdomen and smart, banded legs.

These spiders settle in cracks of weatherboards, holes in brickwork and joinery, or gaps around windows and sliding doors. When the tunnel hideaway is lined with silk, the occupier extends its patch, covering the walls of the house, or even the corners of the windows and doors, where the spiders cash in on the

attractiveness of illuminated windows to flying insects. There the webs collect dust, debris and the associated ire of humans.

Harvestman

MOST NEW ZEALAND GARDENS host another member of the Class Arachnida — the harvestman (pictured below). Some people refer to them as 'daddy longlegs'. Their legs are indeed very long, thin, feeble, and curved or distorted in such a way that it is hard to imagine how they actually walk with them, especially at the speeds they do.

Despite the fact that we have a number of native species, the European harvestman (*Phalangium opilio*) is the most frequently encountered species. It presumably got its common name because of the huge populations that were found in English fields at harvest time.

Harvestmen have a segmented abdomen and no distinct 'waist' between this body part and the cephalothorax (head-thorax). They have a couple of very simple eyes that can only tell the difference between day and night, which is important for the harvestman as it is prone to drying out. It prefers to be active in the cool of the night. During the day it hides in moist habitats, under dense vegetation, logs and stones.

They are scavengers, preferring to eat all sorts of dead proteinaceous materials — insects seem to be their favourite, but they have also been observed to capture live, small invertebrates or their eggs to eat. Prey and food are detected by the sensitive extremities of the second pair of legs — that's why you can sometimes see them raising these legs as if to probe the road ahead.

Unlike spiders, harvestmen do not produce silk or immobilising poisons, yet they are not easy meat for other predators as they defend themselves by excreting a repugnant fluid that oozes from the body when stressed or threatened.

Jumping spider

MOST SPIDERS PRODUCE SILK in some way or another, although this may not always be obvious. One of the cutest Arachnid critters is the small jumping spider (Family Salticidae; pictured below). Their modus operandi of catching prey is

well known — they simply stalk a suitably sized insect and, literally, jump on it.

If you could slow down the action you would see that the spider anchors a thread of silk to its original perch before it jumps. While it is in mid-air and going for the kill, silk is reeled out at an incredible speed, so that the predator always has a lifeline should it miss its target or fail to make it to the next landing site. Young jumping spiders, which need a bit of practice, use this lifeline to prevent themselves from crashing down as they learn the ropes.

In order to be successful at the incredible stalk-and-jump game, the jumping spider needs good stereoscopic eyesight, not only to assess its prey, but also to accurately judge the jumping distance. When you observe a Salticid at close range you'll notice the large front eyes and the somewhat smaller lateral eyes at the front of the cephalothorax. A further two eyes are situated on the sides of this body segment.

The large eyes literally stare at you and as long as you're within 20 centimetres the spider will have a good stereoscopic view of you. These eyes can move from side to side and up and down to 'scan' the target and increase the field of view. The smaller eyes simply extend the view to a full 360°, allowing a stalking spider to keep an eye on all of its surroundings at the same time.

When a Salticid registers an interesting object at the side or the back of the body, it often turns around to take a good look at it through the front eyes. There is a lot of fun to be had confronting an adult male jumping spider with a mirror. The macho bloke will recognise its reflection as an opponent and will immediately go into display-mode, with leg-raising and side-stepping, or even signalling with the coloured palps — small, jointed appendages at the front of the cephalo-thorax. Communication between these spiders is of a visual nature.

Katipo

THE KATIPO (*LATRODECTUS KATIPO*) is sometimes considered the New Zealand version of the Australian redback, that's how closely related they are. In the field it is rather difficult to tell them apart. Interestingly, we have a second species in New Zealand, the black katipo (*Latrodectus atritus*). This spider lacks the characteristic orange-red stripe on the back and looks a bit like a smaller version of the black widow. It is distributed in the northern half of the North Island.

Katipo spiders live near beaches, and not in peri-domestic situations like the redback. Its preferred habitat is driftwood, marram grass and even discarded drink cans. Their messy web — sometimes labelled a spaceweb — is often found close to the ground. Their prey consists of a wide range of invertebrates, usually beetles and sand hoppers that cruise or jump in the general vicinity.

New Zealanders are most at risk when they hang their beach towels over a log of driftwood or on the marram grass to dry. The spider gets onto the material, and will bite when the towel is later draped over the shoulders. You'll definitely feel the bite. Common-sense beach-towel management

could be a potential trouble-saver.

But there are problems on the horizon for our two native katipo. Their numbers appear to have been dwindling slowly but steadily, and one study pointed the finger at competition from a related but introduced species of 'false katipo', *Steatoda capensis*.

This beast is native to South Africa and was inadvertently introduced to New Zealand. It too makes its messy spacewebs along beaches and dunes — exactly the same habitat that katipo frequent. The hypothesis is that Steatoda is more aggressive, hence taking over the katipo's preferred websites (excuse the pun!).

Steatoda look very similar to adult black katipo spiders, but are a much shinier black. Sometimes a few white spots adorn the pea-sized abdomen, and certain individuals also sport a very narrow orange stripe, right at the abdominal tip. You can find similar members of the Genus Steatoda in your garden, in between rocks, inside the holes in bricks and under garden seats. They

too can deliver a painful bite when provoked, but it isn't packed with the poisonous punch of Latrodectus.

You can always tell a Latrodectus species because it has a bright orange-red hourglass mark on the underside of its abdomen. Steatoda never have such a mark.

Nurseryweb spider

HAVE YOU EVER SEEN those large, dense, white webs at the edges of gorse bushes, or other vegetation? They're unmistakably the creation of the native nurseryweb spider, *Dolomedes minor* (pictured below). It's a beautiful and large species but, unfortunately, seldom seen during the day when it hides below the web, keeping an eye on its genetic investment. At night the female will often crawl over the web to guard it.

Inside the gleaming fortress is the silken egg sac containing the developing spiderlings. These babies have to moult their skins at least once before they start roaming

around their effective and protective nurseryweb. After a week or so they emerge from the web by opening up the outer layer. They might hang around their maternal nest for a while, but slowly they disperse into the big, bad world.

The nurseryweb spider catches prey by pouncing on anything that comes too close. Silk is not used at all for procuring lunch.

Spaceweb spider

ACHAEARANEA VERUCULATA IS ONE of the native spaceweb spiders, belonging to the same Family (Theridiidae) as the katipo. At first sight it is a rather unassuming spider — with a name larger than its body — that lives in messy webs suspended from the timber of balconies, eaves, sheds, and other man-made structures. A closer look reveals a rather pretty species with a variety of subtle colours and almost translucent, banded legs (pictured below).

This spider has gratefully accepted

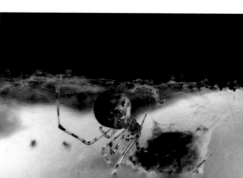

nan-made habitats and is widespread throughout the country. No doubt it takes advantage of the myriad insects attracted to the peri-domestic environment of human households — gardens, compost, manure piles and nocturnal illumination all contribute to the wealth of invertebrate life found hanging around our homes.

Achaearanea suspends her egg sac in the same web she occupies, so that she can keep an eye on the youngsters and perhaps help them on their way with the odd communal meal or two.

Spotted ground swift spider

THE SPOTTED GROUND SWIFT spider (*Supunna picta*; pictured below) is another Aussie interloper that snuck into New Zealand 60 or 70 years ago. It is now widely distributed throughout the country and can be easily identified by its speed — it is the fastest arachnid we have. It has a habit of running around, especially on open ground, for short bursts and then freezing.

Their colouration makes them surprisingly difficult to spot when they sit still. They're predominantly black with smart white markings on the abdomen and cephalothorax. The front legs are orange and these are often held aloft so that the spider resembles an Australian species of solitary hunting-wasp, waving its orange antennae. That's clever stuff — these wasp-mimics rely on the fact that no-one wants to attack a wasp. This enables them to hunt in the light of day.

Supunna belongs to the Family Corinnidae — the family of fleet-footed spiders. It hunts anything on the ground that comes within its visual range.

Vagrant spider

THE VAGRANT SPIDER (*MITURGA* species) on the other hand, hunts at night and belongs to the family of prowling spiders. It's a bulky brute — up to 25 millimetres in body length. Originally confined to the forest floor and tree trunks, some species are quite capable of sustaining populations in heavily modified habitats such as orchards and gardens.

Miturga hunt in a similar way to the Avondale spider and will retire to their little silken shelter sites — under logs, rocks or garden ornaments — when the sun comes up. However, it is not uncommon to see male vagrant spiders inside or near a house. They tend to go out on the town from time to time, especially in autumn, to score themselves a partner. Their journey can lead them into all sorts of strife, including slippery bathtubs and the wrong end of a broom.

But this is nothing compared with the treatment Miturga can get from some of our native spider-hunting wasps. These wasps grab a live spider, paralyse it with some poison, and store the still-live prey next to their own offspring as a living food supply for weeks to come.

The golden spider-hunting wasp roams the soil and vegetation for suitable spider prey. When such a spider has been found and subdued, the wasp will drag it — against all odds and with admirable tenacity — to its own burrow in the soil.

Water spider

A CLOSE RELATIVE OF the nurseryweb spider is the water spider, *Dolomedes aquaticus* (pictured below). Although slightly duller in colour, it is nevertheless a magnificent predator with a body length of about 20 millimetres. Most people assume that spiders do not like water, but the water spider lives up to its name. It lives on the edge of rivers, streams and ponds and shelters under rocks during daylight hours. At night it emerges and literally walks on water with its legs stretched out, without breaking the surface tension of the liquid. The numerous fine hairs on the undersurface of its feet help the spider to stay afloat.

The legs are sensitive to ripples made by potential prey. As soon as an insect comes near, the spider quickly captures it by impaling the hapless victim with its fangs, and then it runs to the shore to devour the catch.

When disturbed, water spiders can run down the side of a rock into the water and surprisingly they keep on going to some depth! They have the amazing ability to trap air between their abdominal hairs — they look as if they have been enveloped by quicksilver — so that they can breathe underwater for prolonged periods of time.

Next time you see a water spider with its diving bell, clinging to a rock underwater, try to dislodge it and it will shoot up to the water's surface, usually ending up upside down. However, quick as a flash, the spider will right itself and scurry off over the surface of the water.

White-tailed spider

WHITE-TAILS CAN BE found right throughout the country, from North Cape to Bluff, and nearly always in close proximity to human habitation. The warmth of human dwellings seems to be perfect for their development. The eggs need a temperature of at least 20°C in order to successfully hatch.

They've been in New Zealand for at least a hundred years. Recently a revision of the Genus Lampona showed that there are two species of white-tailed spider in New Zealand — *Lampona cylindrata* and *Lampona murina*. They are similar in appearance, up to 15 millimetres in length, with a charcoal-grey body and dark rust-brown legs (pictured opposite right). The unmistakable diagnostic feature that sets the Lampona species apart from any other spider here is the distinct, white or light-grey tip of the cylindrical abdomen or tail. Some models, especially the younger versions, also show a few pairs of dark-grey patches on either side of the abdomen.

The white-tailed spider mostly eats other spiders. It is a nocturnal hunter that stalks its prey with great care and agility. They enter the webs of their prey and lure them out of their retreats, often with fatal consequences. Their favourite prey is the grey house spider (*Badumna longinqua*).

Most spiders have some form of poison to inject into prey, and white-tails are no exception. Their poison certainly packs a punch, as many folk who have been bitten by them will confirm. Symptoms include pain around the wound and beyond, followed by swelling and a feeling of general malaise. Sometimes a small blister may form, often associated with a patch of necrotic skin tissue.

If you're lucky, these symptoms will go within a week or so but the after-effects of a white-tail bite can be dramatic — large necrotic areas can develop that take months and months to heal. There is very little direct medical evidence of cause and effect, but most of the circum-stantial evidence suggests that the spider carries on its mandibles a bacterium, *Mycobacterium ulcerans*, picked up from the soil. This organism has also been isolated from the wounds of bite victims and has been proven to cause necrotic lesions on human skin.

Another explanation involves the

SPIDERS AND MITES

spider's digestive juices, which are excreted by the white-tail's mandibles. These juices dissolve the harder tissues of the prey so the spider can eat them. Perhaps the digestive juices also react with human flesh, causing it to break down.

Whatever the result of the on-going research may be, there is as yet no antidote. Ice packs and antihistamines appear to be the first line of defence, and keeping the wound clean and sterile is recommended. When necrosis becomes evident it may pay to get a test done for acid-fast bacilli, which could indicate the involvement of Mycobacterium. In any case it is advisable to seek medical treatment.

How do those white-tails get inside the house? You can often see them cruising the ceiling and walls, especially in spring and summer. They hunt around the exterior of houses at night, and an open window provides easy access.

The problem is that these spiders hide in all sorts of nooks and crannies during the day — a shoe, inside clothes, behind a towel — and the moment a white-tail gets 'trapped' between an item of clothing and human flesh, a bite is inevitable.

Does this mean we should just kill all the white-tailed spiders we encounter around the house? I don't think so. It is more important to be aware of what they look like and their habits so you can avoid getting into strife. After all, these wonderful Australian predators do their sneaky stuff at night, preying on their Aussie mates from way back. What could be more entertaining?

Spider mites

MITES (ACARI) ARE ALSO members of the Arachnida class, and the number, shape, size and lifestyles of these species are truly bewildering. There are so many species, both indoors and outdoors, and some can have an absolutely devastating impact on us, our health, or the health of our crops.

Spider mites are eight-legged invertebrates — that's why they are often referred to as red spiders. Like their relations some spider mites can spin copious amounts of fine silk threads (pictured below) around their feeding sites — a gossamer of silken highways, used for dispersal and as a form of protection from inclement weather. You see, spider mites just love warm and protected conditions.

Because of their size, mites are usually very prone to drying out and therefore require some form of humidity for survival. But the ubiquitous two-spotted spider mite (*Tetranychus urticae*) seems to prefer a dry and sunny outlook from its host plant.

Host plants are almost too numerous to mention. They'll attack just about any plant grown under glass — all the regular vegetables and ornamentals. You'll find

them on roses, especially when these are grown under the eaves on the north side of a house or fence. Remember, out of the rain and in the sun — a classic site.

Life cycle

Although mites are quite different from insects they share a few typical invertebrate characteristics in their development, such as egg stages and skin moults for growth. Similarly, mite reproduction tends to be sexual, sometimes with a distinct difference in appearance between males and females. However, in some species parthenogenesis may be the normal or occasional mode of reproduction.

Eggs are often simple in form and are usually smoothly rounded or oval. Plant-feeding spider mites normally attach their eggs, singly, to the surface of host plants. Development from egg to adult is also simple, with little or no metamorphosis — a miniature version of the adult hatches out from the egg. However, this first nymphal stage possesses only three pairs of legs compared to the adult's usual four pairs.

Most mites undergo three or four moults before they are adult. Just prior to each moult they enter an immobile resting stage in which the legs are tucked up close to the body. All stages feed on plants.

The life cycle of spider mites is almost always short, often taking only a matter of a few weeks under summer conditions, meaning infestations can develop very quickly.

The two-spotted spider mite, which is a small, translucent mite with two dark green spots on either side of its blunt abdominal region, over-winters as red-coloured females in nooks and crannies in bark, litter, and other hiding places such as in the structures of glasshouses, and in the furry patches at the calyx end of kiwifruit. Eggs are laid in spring and numerous generations are completed by the time temperatures drop in autumn.

Impact

A lot of people overlook the Acari, probably because the critters are so small, but quite a few dedicated growers will be all too familiar with the plant-damaging mite species: spider mites, cyclamen mites, bud mites, erineum mites and gall mites.

Those last four groups are usually responsible for distortions and galls on plant material and, although a real problem from time to time, they are not as frequently encountered as the spider mites.

As can be expected, the nature of plant injury depends to a great extent on the type of mite involved and the species and part of the plant attacked. All spider mites are real raspers and hence feed on individual cells, resulting in minute flecks (yellow stippling, sometimes bronzing or silvering; pictured overleaf) appearing on leaves or other plant parts. Most feeding takes place from the undersides of leaves but injury is often seen more readily on the upper surface. Flowers may be attacked as well as foliage. There is usually little or no distortion of the plant parts attacked.

Some species, in particular the two-

SPIDERS AND MITES

spotted mite and the gorse spider mite (*Tetranychus lintearius*) produce considerable amounts of fine silken webbing over the plant surface, but other species of tetranychids, such as the European red mite (*Panonychus ulmi*), do not produce webbing.

The latter damages a range of fruits, especially pipfruit, by mottling and bronzing leaves or reduction of fruit buds for the following year's crop. They are bright red mites that can often be found on the underside of leaves; the eggs over-winter hidden in the calyx of fruit.

The gorse spider mite has been introduced to control this exotic weed and although it sometimes takes off with a spectacular hiss and a roar, it doesn't always kill the gorse bush outright.

Control

Traditionally spider mites have been bombarded with a wide range of chemical miticides: Agrimec (active ingredient avermectin, a by-product made by the soil organism *Streptomyces avermitis*), Apollo (tetrazine), Pentac (dienochlor), Kelthane (dicofol), Peropal (azocyclotin), Torque (fenbutatin), Neoron (bromopropylate) and Omite (propargite) are some of the registered products on the market. Many products previously registered are no longer effective on two-spotted spider mites. It is generally recommended not to use the same product more than three times per growing season to avoid development of resistance.

Note that these materials are all miticides, as the difference between mites and insects is sufficiently large that most insecticides do not affect mites in a lethal way. In fact, insecticides tend to kill the insect predators of spider mites and therefore are of benefit to mites! Having said that, some insecticides will control mites, for example Mavrik (taufluvalinate) and dichlorvos (potassium salts).

Fatty acids will work in an abrasive fashion on the tender mite bodies, especially when applied late in the afternoon, so the plants remain wet for longer periods of time. Mineral oils and neem oil are also pretty good organic miticides.

Cultural and physical control often hinges around altering the environment to deter pests. In the case of two-spotted spider mites, their love for warm and dry conditions could be utilised: mist the susceptible plants with water in the evening to achieve a cooler and wetter environment, but watch the fungal problems associated with prolonged wetting. Raising the relative humidity and giving these two-spots cold, wet feet generally causes populations to crash.

Crop hygiene and removal of crop debris and alternative weed hosts is also of importance.

Predators

Biological control of two-spotted spider mites has been practised right around the world — *Phytoseiulus persimilis* (marketed in New Zealand under the names Mite-E and Spidex) is currently the main thrust of control. This predatory mite is brownish in colour and slightly larger than the two-spots and it will devour all stages of this pestiferous prey as well as other small mites.

Phytoseiulus can be bought 'off-the-shelf'and will happily work outside and in glasshouses. It runs fast over the leaves and stems of spider mite-infested plants in search of eggs, nymphs and even adult prey. It's a gluttonous critter and will kill up to a dozen eggs or spider mites on a sunny day. The development of IPM programmes using Phytoseiulus against spider mites is another strategy to prevent resistance, while gaining good control of the pest.

Amblyseius cucumeris is another commercially available predator, also known as Thripex and Mite-A. It is mainly marketed as a biological control species against western flower thrips, but will also have a go at broad mites, tomato russet mites and two-spotted spider mites. It is quite capable of sustaining itself in the outdoor environment. Amblyseius mites require pollen for additional nourishment, so it pays to companion-plant veggies and flowers to suit this particular predator.

Some predatory insects are also helpful. A small, black ladybird, *Stethorus bifidus*, can be found around pipfruit orchards. It is the apple-grower's mate as its exclusive target is the apple spider mite. *Stethorus bifidus*, and its relative on citrus, *Stethorus histrio*, feed mainly on plant-damaging mites (two-spotted spider mite, European red mite and citrus red mite) and leave the predatory mites alone. Now that's what I call a co-operative predator! Both adults and larvae of Stethorus feed on mites, and large numbers can be found when you have a major mite problem.

ITE INFESTATIONS

THOUGH AN INSECT INFESTATION IN STORED FOODS IS NOT USUALLY DETRIMENTAL TO MAN HEALTH, SOME MITE SPECIES CAN CAUSE A FEW PATHOLOGICAL PROBLEMS. FLOUR TES AND MOULD MITES ARE VERY COMMON IN PANTRIES, ESPECIALLY WHEN THE STORED OD IS GETTING PAST ITS USE-BY DATE. MILLIONS OF THESE TINY, SLOW MOVING BEASTS N SOMETIMES BE FOUND SWARMING OVER THE SHELVES AND WALLS OF THE PANTRY. HE DRIED-FRUIT MITE (*CARPOGLYPHUS LACTIS*) IS QUITE A MEDICAL NUISANCE. THIS ECIES INFESTS SEMI-DRIED FRUIT, USUALLY DATES, FIGS, AND APRICOTS. THEY ARE TEN DIFFICULT TO DETECT INSIDE THESE FRUITS, EVEN WHEN THEY OCCUR IN LARGE MBERS. INGESTION OF THE MITES CAN LEAD TO PULMONARY ACARIASIS, A CONDITION HEREBY THE MITES ENTER THE LUNGS OF THE CONSUMER.

Beetles

Bark beetle

BARK BEETLES HAVE ALWAYS been a bit of a taxonomic fighting ground for entomologists. The most recent way of thinking classifies them as a sub-family of the weevils, which means they have lost their independent 'status' of bark beetle Family Scolytidae.

That is a pity, because I have always found bark beetles to be cool characters. They truly are the masters of artful tunnelling and beautiful patterns. There are a number of species, most of them introduced from Europe or Eurasia.

Life cycle

The dead giveaway of bark beetle activity is the fantastic and regular pattern of brood galleries, formed under the bark (pictured below). The female beetle cuts a long gallery, often parallel to the grain of the timber. At regular intervals she cuts shallow egg-laying bays on either side of this main gallery. In each bay she places a single egg.

The rest of the pattern is completed by the larvae that hatch from the eggs. They tunnel more or less at right angles away from their mother's gallery. As the larvae grow, they chew a wider tunnel to accommodate their plump bodies.

Having constructed such a nice, protective tunnel as a larval project, it only makes sense that pupation takes place at the end of this journey through the wood. When the next-generation beetles emerge from the pupal cases, they chew their way out through the bark, find a mate, and start the whole life cycle process over again.

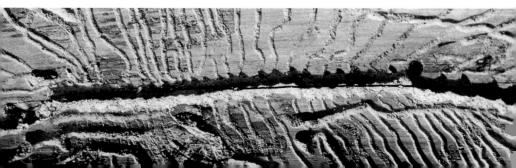

Impact

Bark beetles are not necessarily a lethal invertebrate, as they have a habit of targeting trees that are already under stress or logs that have recently been felled. They particularly like timber lying in a damp, shady place or in contact with the soil.

Bark beetles often carry with them fungi that aid the larval development. Sapstain fungus is one example; Dutch elm disease is another, rather famous, fungus that gets distributed by flying bark beetles. The latter is usually fatal to its host.

It is just about impossible to control the bark beetle species that damage trees. Even if you were able to *predict* which tree(s) they are going to go for, there is very little you can do to prevent their access. In the case of Dutch Elm disease the strategy has to be to cut and burn affected trees as soon as possible to stop the beetles spreading, and hence spreading the disease.

Prevention of bark beetle damage to recently felled timber is achieved by common sense sylvicultural practices such as rapid processing and removal of felled logs.

Carpet beetle

THERE IS NO DOUBT that the carpet beetle is a pest. In New Zealand we have two species of carpet beetle: the Australian carpet beetle (*Anthrenocerus australis*) and the more colourful and common varied carpet beetle (*Anthrenus verbasci*; pictured below), but their preferences and habits appear to be more or less the same.

Adult carpet beetles are shaped a little bit like miniature ladybirds, measuring 2 or 3 millimetres from head to tail. The Australian species is almost uniform black with smudgy light-coloured patches. The varied carpet beetle is quite beautiful as its wing covers are clothed with colourful scales — white, yellow, rusty-brown and black are the predominant hues.

Ecologically speaking, the carpet beetle has been designed by the Almighty to perform a unique task which very few other animals are able to tackle — it can digest fur and wool. It does this by breaking down the sulphur bonds in the keratin molecule. This means that a dead furry animal can be completely recycled into an assortment of chemical elements, most of which, in the long run, benefit our plants. In areas where

BEETLES

carpet beetles do not occur, cadavers will often survive for prolonged periods as flat bits of skin and hair, months after all the meat has been processed.

Life cycle

Adult carpet beetles fly around our gardens in spring and are very partial to a diet of pollen and perhaps even nectar. In view of this, these beetles may play an important role in pollination. After all this wonderful pollen feeding, the fully developed beetles will be on the lookout for a mate.

Fertilised females, on the hunt for a nice piece of 'flat skin-and-hair' to lay eggs on, often fly straight indoors, where they find many square metres of dead sheep lying on the floor, just waiting to be recycled. More traditional oviposition sites are dead, dry animal or bird carcasses, or even a bird's nest, where the shed-skin debris and feathers form a good larval substrate.

Indoors, the stripy larvae (pictured below) appear in late summer and take their time chewing through the carpet. There are a few skin changes, or moults, before they pupate in early springtime.

Adult beetles emerge a few weeks later and are often found dead on the windowsill in September–October. They were on their way out to the garden when they bumped into that unforgiving window pane.

Impact

In the winter months you'll find the larval forms devouring carpets and woollen clothes. The grubs, which are a few millimetres long, resemble transversely striped, hairy maggots and are often encountered deep in the pile and on the underside of carpets, especially around the

edges or in damp, undisturbed places, such as under the piano or other heavy furniture, where a predatory vacuum cleaner is rarely seen.

Have you ever seen the damage these creatures can do to a valuable carpet? Often, the first sign is the vacuum cleaner sucking up a heck of a lot more than just dust — whole patches of woollen carpet pile disappear, leaving behind bare areas of synthetic backing. In amongst the debris you'll find fragments of striped carpet beetle skins and crawling, striped larvae. It's a mess, and it's an expensive one to fix.

When you get your woollen sweaters and cardigans out of their summer storage drawers, take a look at their condition. Holes in the woollen fabric indicate the activity of larval keratin feeders, usually clothes moths, but also carpet beetles. Caterpillars, moths or beetle grubs won't be difficult to find nearby.

Early winter is a good time of year to look for signs, as the larval part of the life cycle will be in full swing. The beetles seem to prefer carpet areas with a higher-than-average relative humidity and little disturbance — look along walls, under beds and furniture, and near large floor-to-ceiling windows on the south side of the house.

Control

As mentioned before, in spring and summer you'll find dozens and dozens of these beetles feasting on the pollen of some important plants — Queen Anne's lace certainly is a popular host, as are daisies.

I've always advocated the use of Queen Anne's lace to attract beneficial predators and parasites to help with biological control of garden pests, but they obviously attract all sorts of other insects, so keep an eye on your plants and be forewarned.

Carpet beetles are relatively easily controlled with residual synthetic pyrethroid insecticides — materials such as permethrin, cypermethrin and bifenthrin will protect any furry food source from the gluttonous activities of the larvae. But, the trick is to find them before they start chewing.

Chafer beetle and bronze beetle

Life cycle

AS THE FORCES OF Nature synchronise their watches (governed by day-length and spring temperatures), chrysalises and pupae hatch into adult insects that waste no time preparing for a successful — shall we say — orgy. And chafer beetles (pictured overleaf at left) and bronze beetles know exactly how to party.

The first thing these newly hatched beetles do is gather in swarms of hundreds, sometimes thousands, of individuals. Mind you, as this is done under the cover of darkness we tend to be blissfully unaware.

Then they set off in search of a good-tasting food source — anything will do — cherries, plums, elms, camellias and even succulents such as jade plants. The idea is to devour as much plant material as possible in order to develop the stamina required for reproduction.

If you keep an eye on susceptible trees and shrubs, when you discover the first signs of shredded foliage in the morning, all you need to do is look under the affected plant. In the vegetation you'll see dozens and dozens of beetles doing their thing in the undergrowth.

The fertilised females usually do not move too far away and will deposit their eggs in cracks and crevices in the soil, so that the offspring will be right in their favourite element when they hatch from the eggs.

The creamy-white grubs (pictured below right) dig downwards and chew for their entire larval life on the roots of the plant, shrub or tree on which the parents fed in spring. Thank goodness a lot of gardeners are not aware of this double whammy damage; what we don't see or don't know about is usually no real worry.

In the greater scheme of things I find it appropriate that these beetle species cause chewing damage on the same tree both above and below ground level. It fits nicely with the delicate balance that exists between leaves and branches versus root biomass.

Impact

Spring is the season of disappearing leaves and the damage after an attack by a horde of native yellow-spotted chafers (*Odontria xanthosticta*) — or bronze beetles (*Eucolaspis brunneus*), for that matter — is likely to be serious indeed, with ragged and shredded leaves on ornamental and fruit trees alike: damage so severe you'd swear it was caused by possums. The problem is that this can occur at any time in spring, from September well into December.

Control

If you could predict which tree the beetles were going to attack in spring, and when, you could deposit a persistent insecticide on the leaves. But you can't, and therefore you would be wasting your time and money. Besides, the chewing damage might look awful, but it hardly rates on the scale of plant health.

Grass grub

AFTER SUMMER, WHEN THE soil's
softened up with the autumn rains, take
a look below the surface of a patch of grass
at the wonderful world of the grass grub
(*Costelytra zealandica*).

We've got a real character here —
a native beetle species that has been eking
out a living here for millions of years, long
before *Homo sapiens* arrived on the shores
of Aotearoa. As with so many native
insects they were minding their own
business, nibbling on tussocks and generally
enjoying the light, free-draining soils, up
to about 1200 metres altitude. And then
the pastoralists arrived.

It has been said a thousand times: the
backbone of New Zealand's economic
status is its pasture and the production of
agricultural and dairy products. When
you take a look back in our recent history a
heck of a lot of bush and rough terrain has
been 'improved' and turned into pasture.

Grass grubs have adapted themselves
brilliantly to the new grass species. Ryegrass
is one of their favourite items to feed on.
And on top of all that, the adult beetles
love to chew brassicas and pasture plants,
but they are especially keen on the fresh,
young foliage of a wide range of
introduced deciduous trees, or small,
developing fruits such as apples. I sometimes wonder if the term 'improved
pasture' refers to it from the grass grub's
point of view.

Life cycle

The adult beetles wake from their pupae
in spring when the warmer temperatures
kick-start them into action. From October
to mid-December the brown, 10
millimetre-long adults will be on the wing,
frequently in large numbers. They are
attracted by bright outdoor lights and the
swarms can be a dead giveaway. In warmer
districts the hatch will be around October–
November, whereas further south activities
peak November–December.

As with most insects, the adult beetles
are the reproductive stage of the life cycle.
Males find females by the pheromones that
these ladies emit when they hatch from
the soil. As a result, the male beetles
(pictured left) fly at low altitudes in order
to track down their target.

But before the males look for a
partner they need to feed, so that their
genitalia are in top condition. Adult grass
grub beetles devour a wide range of
foliage and fruits — the raised brown
lumps commonly found on apple skins
are often the healed scars of damage
caused by adult beetles feeding in spring
when the fruit is very small.

No doubt females also participate in a
bit of foliar feeding before they lay their

BEETLES

clusters of eggs in the soil. The beetles are equipped with coarsely serrated front tibiae that make excellent digging devices and a female thinks nothing of digging 200 millimetres into the soil in order to deposit her dozens and dozens of eggs.

Three weeks later the first instar larvae hatch and these waste no time starting to feed on the roots of host plants. They are small, dirty-white, C-shaped grubs with a light brown head (pictured below right). They show surprising agility in

nice, loose soil conditions. The first instar grubs are around for about three months (November– January) after which a skin moult occurs.

They continue feeding on the roots and will do so for about six weeks, moving up through the soil as plants die off, in search of fresh fodder. After the next moult in March the fat, plump third instar larvae take over the job of severing the roots from the plants in the top layers of the soil (about 25 millimetres deep).

When fully fed, at the onset of winter in June, the larvae move to deeper soil levels to over-winter. They may go as deep as 250 millimetres. A period of procrastination follows during the winter months (entomologists call this stage the pre-pupa), during which the larvae empty their gut in preparation for pupation in October. The pupal stage (pictured below left) lasts about five weeks and takes us to where it all started a year earlier — the spring flights of adult grass grub beetles.

In some cold southern climes and on cooler hill country the larvae can take two years to complete a full life cycle, and when there is a major drought or a distinct lack of food, larvae can also take time out and switch to a two-year life cycle.

NORTHERN GRUBS

MOST PEOPLE IN THE WAIKATO AND POINTS FURTHER SOUTH WILL KNOW WHAT GRASS GRUB DAMAGE IS ALL ABOUT. IN THE NORTHERN PARTS OF THE NORTH ISLAND THIS SPECIES RARELY CAUSES ANY DAMAGE. THERE, THE BLACK BEETLE (*HETERONYCHUS ARATOR*) AND THE EQUALLY BLACK TASMANIAN GRASS GRUB (*APHODIUS TASMANIAE*) TEND TO TAKE THE GRASS GRUB'S PLACE. THEY ARE ALL RELATED SPECIES, BELONGING TO THE SAME BEETLE FAMILY SCARABAEIDAE, WHICH ALSO INCLUDES CHAFER BEETLES AND DUNG BEETLES.

Impact

Pasture damage is often severe — brown circular or irregular patches of dead grass that can be rolled back as if it were Readylawn. These bare patches of soil are prone to invasion by weeds.

Control

Where are the weak spots in this well-designed insect's armour that allow for some serious control measures?

Well, first of all, trials with grass grub pheromone in traps have shown that it is actually possible to accurately pinpoint the peak flight nights in spring. The paddocks, complete with copulating adults, can then be treated with residual insecticides before large numbers of eggs are deposited in the soil. Unfortunately, this technique is still cost-prohibitive. Insecticides, expensive as they may be, are still used, mainly in autumn when larval densities get to around 100 larvae per square metre. A number of formulations active in soil are registered for grass grub control, but the wily old grub has shown signs of resistance to certain active ingredients.

When we look at the more organic options, the bacterial preparation Invade (*Serratia entomophila*) causes mortality in larvae and it remains present in pastures for several years, giving good biological control. Unfortunately, this natural 'amber disease' of grubs is not yet available for the home gardener, but the future may look bright in this respect.

There are other tricks we can employ

to minimise populations of this destructive grub. A single cultivation in spring successfully damages the fragile pupae, but in autumn cultivation needs to be repeated a few times in order to destroy significant numbers of larvae. Heavy grazing of pasture by stock in summer has also been shown to be beneficial, as it reduces the growth of grass plants, and hence minimises the amount of food available to the grubs. An added bonus is that in very short pasture the summer sun heats the soil to temperatures lethal to grass grubs during prolonged dry spells.

Pasture composition also plays an important role in the wellbeing of grass grub. Ryegrass and white clover are the ice cream of this insect; they love it. Lucerne, lotus and phalaris are resistant pasture species that actually suppress grub numbers, and tolerant species that withstand the insect's nibbling include tall fescue and cocksfoot.

Natural grass grub disasters make a difference — severe droughts in summer kill eggs and first instar larvae, and similarly, very wet winters take their toll on the over-wintering larvae and pre-pupae in sodden soils.

And speaking of sodden soils and grass grub disasters — remember the third instar larvae that dwell in the top layers of the soil from March to June? If you have a really wet autumn, with squishy-wet soils, why not borrow all the stock from your neighbour's place and pen them up over the grass grub infested area? The resultant downward pressure of myriad hooves will squash the offending insects. It's a Kiwi invention called 'mob-stocking' and a four-legged variant of my favourite pest-control method — digital control.

If you are looking at controlling grass grubs in a city lawn where a herd of cattle would probably look out of place, you can always try the townie version of mob-stocking. It involves a very heavy, water-filled roller and a good work-out.

Predators

Naturally, certain predators (starlings and rooks) love to feast on the grubs, but their effect is probably not measurable.

Huhu beetle

SAY WHAT YOU LIKE about New Zealand, nothing beats a summer evening in Aotearoa, no matter where you are — Central Otago, Marlborough, Hawke's Bay or north of the Bombays. On such a night the whole entomological circus performs, triggered by the lights in your lounge, which causes confusion to a veritable army of flying insects. One of the biggest has to be the huhu beetle (*Prionoplus reticularis*; pictured below).

Huhu belong to the longhorn beetle Family Cerambycidae, which makes sense when you see the length of their antennae. They are up to 50 millimetres long — and they bite and have sharp claws. When you take a close look at them (when they finally sit still) you've got to admit the huhu beetle is a magnificent insect.

The wing covers or elytra are dark brown, hard shields adorned with slightly raised, lighter-coloured veins in a reticulate pattern. Underneath these tough, protective elytra are the large, translucent flight wings which can be folded up neatly and tucked away out of sight.

The thorax, abdomen and parts of the head are clothed in golden brown hairs, giving the beetle a furry look, perhaps in an attempt to appear a bit more endearing.

Don't be fooled by all this fur. Even if you're an entomologist and know how to pick up such a beetle (holding the thorax with thumb and forefinger) you could be in

or a prickly surprise — there are a few pretty sharp spines on that body segment!

For some reason the huhu will often wriggle itself free of the grasp and as its mandibles are sharp and powerful, will certainly have a go at human flesh.

Life cycle

Huhu grubs are reared on a diet of dead and sometimes slightly decayed wood. When a tree falls over in the forest and lies there for a couple of months, adult huhu beetles start to get interested in it as a long, cylindrical habitat.

Gravid females lay their cigar-shaped eggs under the bark or in nooks and crannies so that, three weeks later, the small creamy larvae have direct access to the dead cellulose of the xylem. Those small, soft-bodied grubs are often the first inverte-brates to gain access to the moist timber after the death of a tree. As such they blaze a trail in the form of a tunnel for other wood-decaying organisms to follow.

Huhu larvae have been at it now for many millions of years and their ecological role could be described as 'aiding the recycling process'.

Once the larvae are established inside a dead log they chew away, often following the grain of sapwood or heartwood. They usually pack their coarse sawdust excrement behind them in the tunnel, so it doesn't get too messy at the site of chewing activity.

The larval stage of the life cycle can be rather long. It depends a bit on the quality and quantity of suitable wood, but three years of a tunnelling lifestyle is not uncommon for our huhu.

Just before it is time to pupate, a grub can measure up to 75 millimetres in length (pictured below) — that's a lot of concentrated insect protein, contained in a flexible but tough skin. A huhu grub typically shows quite constricted body segments, making it look a little bit like the Michelin Man.

In spring, when it is time to make the final larval moult and turn into a pupa, the larva heads for the surface of the wooden log; there it will excavate a spacious chamber packed and lined with shredded wood fibres for a comfortable mattress. The pupa stage lasts for a few

weeks — then the beetles waste no time emerging, chewing their way out of the log. When the elytra and wings are hardened off, the maiden flight is next on the agenda . . . that's when we often encounter these majestic insects.

Impact

Unfortunately, the huhu sometimes starts its noble job on fresh logs, destined to be cut up for timber, or even on moist, rough-sawn boards stacked in the timber yard.

Control

In the sylvicultural industry it pays to look out for this species and process logs as soon as possible to minimise the window of opportunity for this polyphagous insect. Yes indeed, it eats many different hosts, from commercial pine species to a range o natives.

Ladybird

LADYBIRD BEETLES ARE THE adult form of the life cycle and they come in many brilliant colours. The archetypal red or orange and black ladybird is often depicted in children's books. Yet these colours make the beetles quite easy to detect by birds and other would-be enemies.

Red and black are also warning

TASTES LIKE CHICKEN

IF YOU VISIT THE WILDFOODS FESTIVAL IN HOKITIKA AND TALK TO THE LOCALS ABOUT *PRIONOPLUS RETICULARIS*, CHANCES ARE THEY WOULDN'T HAVE A CLUE WHAT YOU'RE TALKING ABOUT. BUT MENTION 'HUHU' AND THE KNIVES AND FORKS COME OUT.

MAORI HAVE KNOWN OF THE FOOD VALUE OF HUHU GRUBS FOR QUITE A FEW CENTURIES. HUHU GRUBS ARE VERY HEALTHY TUCKER — THEY'RE PURE PROTEIN AND VERY LOW IN FAT. NOT A GRAM OF CHOLESTEROL IN THEM! WHEN A GRUB IS PREPARING FOR PUPATION, IT WILL CEASE FEEDING FOR A FEW DAYS OR SO. AT THAT TIME, THE INSECT WILL HAVE FULLY EVACUATED ITS GUT CONTENT, WHICH MAKES THEM TASTE A LOT BETTER!

MOST PEOPLE DESCRIBE THEM AS 'TASTING LIKE CHICKEN', SOMETIMES WITH SLIGHTLY NUTTY OVERTONES. TO ME THAT MEANS THAT THERE REALLY ISN'T MUCH FLAVOUR AT ALL; IT IS A NEUTRAL FOOD WITH A CHICKEN TEXTURE, PERHAPS.

THAT MAKES IT EXCELLENT AS A CARRIER FOR ANY FLAVOUR YOU ADD: MALAYSIAN, INDIAN, CHINESE, THAI — THE POSSIBILITIES ARE ENDLESS. HUHU GORENG OR HUHU VINDALOO, ANYONE?

Life cycle

While most people are perfectly able to identify the adult form of ladybird beetles, the immature stages, or larvae, are often overlooked. This is no surprise, as they look nothing like their parents in general shape, colour, or even movement.

Ladybird larvae are rubbery critters with distinct body segments, a large, robust thorax and head and a tapering abdominal section (pictured overleaf right). Some are adorned with soft spines. The immatures move about and gobble up anything they encounter, just like gluttonous kids. They'll need to do that in order to put on weight and grow — and of course growth can only be achieved by moulting the skin from time to time.

There seems to be no evidence that they detect their prey by sensory wizardry — it's all a matter of chance. The more or faster they move, the more chance they have of running into lunch! And it helps if mum has deposited her eggs in amongst a veritable smorgasbord.

When larval development has been completed, the ladybird larva needs to change its skin one more time in order to turn into a pupa. After 'gluing' the abdominal tip onto a firm substrate, such as a leaf, some species perform the most exhausting callisthenics (leg stretches on tippy-toes) to get the last larval skin to split. The resulting pupa is always firmly anchored by its last abdominal segment. Just before hatching, some female pupae are visited by male adult ladybirds in an attempt to prematurely mate with them.

colours and ladybirds do not bluff when they use them. The blood and body fluids of both the larvae and adults are toxic to most mammals and birds.

The black colour (melanin) is derived from your regular, run-of-the-mill chemicals (tyrosines) that most insects use to harden their skin. The red colour is synthesised from carotene, which is obtained from plants via their plant-feeding insect prey — think of the red in tomatoes and carrots. When carotene oxidises, the colour turns to orange or even yellow (xanthophylls).

Impact and benefits

It's interesting to note that when we are talking pest populations of aphids on roses and other desirable plants, most New Zealanders immediately mention ladybirds as suitable predators.

One of the most common ladybird species, especially in the South Island, is the two-spotted variety (*Adalia bipunctata*). This European import is a well-known general predator of aphids, but despite the fact that it can reach good population densities in our gardens, it is not a very effective control agent. In spring and autumn aphids reproduce at such an alarming rate that the poor old ladybird has great trouble keeping up with the population explosion of its prey.

Unfortunately, the same goes for the eleven-spotted ladybird (*Coccinella undecimpunctata*), which is another deliberate introduction from Europe. This species lays its eggs on the soil under aphid-infested plants and its larvae can be very numerous indeed — up to 100 larvae per square metre. These voracious eating-machines will chew on a whole range of small insect prey, including caterpillars and even their own siblings. They are true 'generalists'.

Another species with a good, diverse menu is the light orange and black *Harmonia conformis* (sometimes called the eighteen-spotted ladybird). It is known to grab psyllids, especially on Pittosporum, and the odd mealybug perhaps. Oh yes, and if there's an aphid in the way, it might end up as dessert.

A really small, black species, *Stethorus bifidus*, can be found in pipfruit orchards, where it feeds exclusively on the apple spider mite. As long as you don't use insect-icides in the orchard (ladybirds are very

BIG APPETITE

LADYBIRDS ARE INSECTS WITH LARGE APPETITES. THERE ARE PLENTY OF STATISTICS TO SHOW THAT THE NUMBERS OF PREY ITEMS THEY DESPATCH RUNS INTO THE MANY HUNDREDS — 800 APHIDS FOR AN ADULT BEETLE DURING A 40-DAY LIFE SPAN, AND 200 APHIDS FOR A DEVELOPING LARVA. GLUTTONY IS THEIR MIDDLE NAME, AND MOULTING IS MERELY AN IRRITATING NECESSITY OF LIFE.

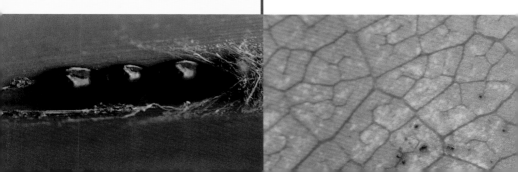

susceptible to these compounds), you'll get the benefits of this small beetle. Keep in mind the general rule about the efficiency of predators: specialists tend to be better for biological control than generalists.

Another predominantly black ladybird (but with a rusty-red head and thorax) is the Australian mealybug ladybird (*Cryptolaemus montrouzieri*). Here's another predator that specialises in one type of prey and it seems to be doing such a good job that it is now being reared in captivity with a view to marketing it as a biological control agent. The pure white and woolly larvae look just like an overgrown version of their mealybug prey and are, therefore, often called 'wolves in sheep's clothing'.

So far we've dealt with ladybird species that are black, red, orange, yellow, or combinations thereof. The next one is a totally different kettle of fish. Iridescent blue or blue-green, *Halmus chalybeus* is the Australian steelblue ladybird (pictured below far left). It derives its colour from a scattering of light that passes through the thin, transparent surface layer of its wing covers or elytra. When this light is reflected back, the yellow is filtered out, leaving the blue to turquoise hues.

The steelblue ladybird is often found on citrus, flax, feijoa, bamboo, hydrangea and other plants, where it cruises the leaves and stems in search of its favourite tucker: scale insects, although aphids may also be consumed. As with most ladybirds, it is the adult beetles that survive the winter in all sorts of nooks and crannies, often in the company of mates.

Illeis galbula is a beautiful black-and-yellow ladybird that has favoured the northern parts of our country for the past two decades. It moves gracefully over your courgettes, pumpkins and other Cucurbitaceae, clematis, oak leaves, dahlias, hydrangeas — in fact, just about any plant that shows a good dusting of powdery mildew on its leaves. This is the mildew ladybird, the only species of ladybird in New Zealand that isn't a carnivore.

It tends to be particularly numerous in late summer and autumn, when mildews thrive in the warm, hot days and cool nights. The resulting nocturnal dew will keep the susceptible leaves nice and wet, and powdery mildew takes full advantage of that phenomenon. If you look around in spring you'll find the adults of this ladybird species in low numbers on mildewy plants — they're hanging in there until autumn, when

population levels rocket skywards.

Illeis galbula is a fashion-conscious species, whereby the adults, larvae and pupae are colour co-ordinated. They both show copious amounts of yellow and black, a colour combination that, surprisingly, affords quite a bit of visual protection on the underside of a mildew-infested pumpkin leaf.

The larvae look smart with their black spots and yellow thorax, as they graze their fungal hyphae. They don't need to move fast. Once they've found their fields of plenty it's a matter of head down and keep on chomping.

People are often surprised to find out that this beetle is a vegetarian or, more precisely, a fungivore and, for one brief moment, they look excited — surely these ladybird beetles are useful for the biological control of mildews?

Sadly, observing these elegant creatures in my veggie garden I think I have gathered enough evidence to accuse them of spreading the fungus from leaf to leaf and plant to plant. You see, these guys are just like us — they cultivate their gardens!

Lemon tree borer

THE LEMON TREE BORER (*Oemona hirta*; pictured above) is a ubiquitous little blighter that causes a lot of strife in people's gardens and crops. It belongs to the Family

of longhorn beetles (Cerambycidae), a group that can be identified by the long, segmented antennae that adorn the adult beetle's head. They are quite pretty, medium-sized beetles, 15–25 millimetres long, and of slender build. The dark exoskeleton is clothed with numerous fine, orange hairs, especially on the wing covers. The legs are dark brown to red-brown. Some orange patches adorn the back of the head and the base of the wing covers, and the thorax (the segment behind the head) is distinctly ridged.

The beetles tend to be on the wing in the early morning and evening, when their living and loving takes place, but sometimes they enter houses, attracted by artificial light sources. When you pick up one of these beetles, you may be surprised to find that it makes a faint squeaky sound.

Life cycle

The adults live for up to two months and peak flight months are October, November and December, but adults can be found any time between September and February. This is the reproductive stage of their life cycle.

The pregnant female beetles 'sniff out' damaged plant tissue. Of course, these beetles do not have a nose — they pick up scents via their long, sensitive antennae. A female can detect cracks in bark, fresh pruning cuts, and scars (where twigs and branchlets have broken off) from kilometres away, probably because of the sap that the tree exudes to heal the wound. These cracks, scars and cuts are the best places to lay eggs. All the hatching larvae have to do is tunnel straight in and start feeding. And that's indeed what happens.

The larvae (even the very small ones) are equipped with diamond-hard mandibles, and they waste no time commencing the excavation of their larval tunnel. These grubs (pictured below) really move, which is surprising as they have no legs to speak of, but the prominent ridges on the body of the grub enable movement and considerable forward pressure.

Every few centimetres the grub constructs a side-tunnel to the outside of the branch, creating a neat, oval hole in the bark. All excess frass (droppings that look like sawdust) is pushed into these side-tunnels, and out through the oval 'toilet' holes (pictured below to the right of the grub).

The larva spends almost two years in its tunnel. Feeding activity slows considerably during the winter months. When the grub is full-grown — about 35 millimetres long — it constructs itself a pupal cell, cushioned with nice, cosy sawdust for insulation, and turns into a creamy-yellow pupa (pictured overleaf). This usually occurs around June.

The pupa of the lemon tree borer is remarkable, as you can pick out some features of the adult beetle formed inside. The long antennae have their own little thread-like containers for development, as do the individual legs and wings. The shape of the head also stands out.

The adult beetle that hatches from the pupal skin stays inside the cell for a number of days, to harden its skin. By springtime (September–October) the beetle pushes its way out of the branch and into the world.

Impact

In the larval stages, Cerambycid beetles have an appetite for cellulose (wood). Some species feed on dead wood — huhu grubs are a good example — and their activity helps with the decomposition of timber, by opening it up for fungi and a whole range of composting organisms. Other Cerambycids feed exclusively on live cellulose tissue, causing damage to a plant's xylem, which affects water and nutrient distribution within the plant.

Our lemon tree borer belongs to the latter bunch of vandals and most people will have seen the results of their larval ravagings in citrus, elms, grapes, Prunus, wineberry, Robinia, poplar, gorse, willow, walnut, gooseberry, wisteria, fig and tamarillo. The list is almost endless. But, as its common name suggests, it has a great liking for citrus trees, especially lemons. It's found throughout the country, so most people that grow woody plants will, at some stage, see the effects of this beetle.

BECOMING A PEST

OEMONA HIRTA MUST HAVE BEEN AROUND FOR MILLIONS OF YEARS IN THE NATIVE BUSH. IT'S AN INSECT ENDEMIC TO NEW ZEALAND, WHICH MEANS IT OCCURS NOWHERE ELSE IN THE WORLD. ITS ORIGINAL PREFERRED FOOD WAS MAHOE, BUT CITRUS IS MORE APPEALING THESE DAYS. THE FACT THAT THIS INSECT READILY ADAPTED TO NEW, INTRODUCED PLANT MATERIAL CLEARLY SHOWS THE IMPORTANCE OF QUARANTINE. THIS INSECT IS A QUARANTINE PEST FOR ALL OUR TRADING PARTNERS. IT IS ALSO A GOOD EXAMPLE OF HOW A HARMLESS, NATIVE INSECT CAN BECOME AN ECONOMIC PROBLEM ON NEW, INTRODUCED PLANT SPECIES.

Control

As with most pests, an understanding of the life cycle is important if you are thinking about control. What do you do when you find a branch, vine or twig with the unmistakable signs of lemon tree

borer grub activity? Well, that depends on the time of year. For most of the warmer months it pays to not prune the affected wood off, as a searching adult female beetle will quickly find your pruning cut and lay her eggs.

However, you can safely prune back infested plant material in the middle of winter when there are no lemon tree borer beetles about and re-infestation is unlikely to occur. From March to about July is a good time, and don't forget to destroy the infested branches, preferably by burning. It also pays to put a pruning paste on the wound, just to make it a bit more difficult for the female beetles to find an oviposition site in spring.

In the warmer months of the year control is a bit more laborious, but can still be a very satisfying job. A spray with a systemic insecticide such as Orthene is not going to have the desired results, simply because the active ingredient doesn't get into the sap stream of the xylem in sufficient quantities to kill the grubs.

To find the exact location of the culprit, you will need to use the visual cues produced by the working larva: piles of frass. Find the freshest heap of sawdust hanging from the branch, wipe it away and you'll find the oval toilet hole closest to the derrière of the offending grub.

Now you have many options. A simple hypodermic needle filled with insecticide will flood the tunnel with lethal fluid. (Some folk insist that kerosene will do the trick as well, and indeed it may. The fumes will kill the insect, but in some host plants the kerosene might do a bit of phytotoxic damage as well, so be careful.)

Another beaut technique involves an aerosol can fitted with a fine straw-like tube in the nozzle. (Kiwicare makes one marketed for use against borer, with permethrin as the active ingredient.) The grub will have no chance of escaping the spray forced into its tunnel under pressure! After spraying, it will pay to plug the hole with some putty to keep the lethal stuff inside the branch for a while.

If you want to go 'organic' and practise a rather perverse form of digital control you'll need a piano wire or a guitar string. Insert the string into the hole and spear the little blighter right up its backside.

Parasite

The native wasp (*Xanthocryptus novozealandicus*) is up to 15 millimetres long. It is predominantly black with delicate white rings around the abdominal segments, white spots on thorax and head, and orange and black legs. The black antennae have a smart white section. It parasitises a range of the wood-boring grubs of longhorn beetles, including the lemon tree borer. Female wasps detect the host borer's damage visually, but the exact location of the grub in its tunnel is determined after careful antennal activity. The wasp then inserts its ovipositor straight through the wood and into the hapless grub.

Flies and mosquitoes

Blowfly

ARE YOU SICK TO death of blowflies spoiling your al fresco evening meals? Are you tired of lying on the beach with blowies landing on your warm, suntanned body? Are your woolly sheep suffering from blowfly strike? And why do blowflies always hover around the dog's droppings in the backyard before they visit the barbecue without washing their six feet?

There are dozens of species of blowfly in New Zealand (pictured below); some are native to our country and others are introduced species that arrived here one way or another. Most of our nuisance flies are recyclers of organic materials of a, shall we say, proteinaceous origin.

Imagine you're driving along a country road at night and a possum is caught in your headlights — KERBANG! Who do you think will be first on the scene? I bet you that, first thing in the morning, it'll be a gravid female blowfly. She has an incredibly sensitive sense of smell, which enables her to find freshly dead meat from miles away. The moment the lights go out in a warm-blooded animal, bacteria start working on the skin, sending signals of decay to all those animals that can be loosely described as belonging to the 'Undertaker Squad'.

Life cycle

A fertilised female blowfly has a limited amount of time to lay her eggs before they hatch into meat-loving and hungry maggots, so a fresh possum carcass will do nicely. The blowfly will circle the cadaver and land in soft, moist places to do some ovipositing. Ears, eyes, nostrils and half-

pen mouth — as well as other orifices that will remain unmentioned — are perfect birthplaces for the blowfly's eggs. Once hatched, the tiny maggots (pictured overleaf at left) don't waste any time before burrowing. Basically, these larvae open up the body for further decomposition. It may not be a great topic for conversation around the dinner table, but this sort of recycling is vital in the overall scheme of natural things. You don't want those dead possums lying on the road as miniature speed humps for the next six months, do you?

The blowfly's larvae — that sounds less threatening than maggots — grow as they munch their way through the food, shedding their skin three times before finally turning into a chrysalis-kind-of-thing, called a puparium. Often the fully grown larvae will vacate their food source before this last skin change takes place. That way they can be assured of a nice, dry and safe place, away from rotting flesh and other meat-eating invertebrates, for their metamorphosis.

When the fly is ready to emerge, it inflates a bladder-like sac at the front of its head to put pressure on the thin and brittle skin of its puparium, which then splits open. The brand-new fly clambers out and will spend some hours inflating its wings and hardening off its outer skin.

Recently emerged blowflies do not usually show their bright and shiny colours for a day or so. By then the flies are on the lookout for some nourishment to complete the internal development of their reproductive parts.

Blowflies love to eat nectar and pollen, and are therefore often regarded as good pollinators of flowers and crops. They will also gather juicy liquids that contain protein. I respectfully ask you to think of damp-ish animal droppings, sweat and decaying meat particles.

When blowflies enter our kitchens they have a knack of discovering any interesting food particle that has been dropped or spilled. Somehow, they'll come back to the same site time and time again; odours must play an important role in the fly's guidance systems.

MEMORY TEST

I'VE NOTICED THAT BLOWFLIES HAVE A SPATIAL MEMORY, ALBEIT SHORT-TERM — PERHAPS ONLY A FEW SECONDS. YOU CAN TEST THIS BY CARRYING OUT THE FOLLOWING EXPERIMENT: NEXT TIME A BLOWIE COMES HURTLING THROUGH AN OPEN WINDOW, IMMEDIATELY RAISE YOUR ARMS AND TRY TO SHOO IT OUT. WHAT YOU'LL SEE IS THAT THE FLY KNOWS EXACTLY WHERE IT CAME FROM AND WILL MAKE A 180° TURN TO ESCAPE BACK THROUGH THE OPEN WINDOW. BUT, IF YOU WAIT A COUPLE OF SECONDS, THE FLY WILL HAVE LOST ALL DIRECTION AND WILL CRASH INTO WINDOW PANES OR ESCAPE IN A TOTALLY DIFFERENT DIRECTION.

FLIES AND MOSQUITOES

Impact

Not all blowflies develop in dead animals — some prefer live animals! There are a few species implicated in fly strike of sheep and other domesticated animals around the farm or lifestyle block. *Lucilia cuprina* is probably the most common. It became established in New Zealand from Australia in the early 1980s. It is a rather aggressive pest that lays its eggs on the sweaty skin and wool of livestock. The maggots literally chew away at the skin, not only causing a real mess and a reduction in quality of skin and wool, but also severe stress to the animal and its owner.

Blowflies are as much part of life in New Zealand as are barbecues. When all the food is laid out on the outdoor table, a gravid female — usually a brown blowfly, *Calliphora stygia* — will arrive in search of an oviposition site. She's often in a bit of a hurry, as the time between mating and the eggs hatching is a matter of a few hours or less. Especially in suburbia, where dead possums are rather scarce, it is not uncommon for these blowflies to fly around with dozens and dozens of hungry little maggots in their abdomen.

If you've got a keen eye you'll see that the female does a reconnoitre flight over the table and then circles once more to straighten up the trajectory for the final approach. As she flies over the barbecue table, she 'blows' her tiny larvae in a neat, straight line onto the exposed food like a B-52 bomber.

BREEDING LIKE FLIES

ENTOMOLOGISTS, WITH NOT ENOUGH TO DO, CALCULATE THE SORT OF STATISTICS THAT WOULD NOT LOOK OUT OF PLACE IN THE *GUINESS BOOK OF RECORDS*. A SCIENTIST EARLY LAST CENTURY RECKONED THAT A PAIR OF HOUSEFLIES COMMENCING COPULATORY ACTIVITIES IN EARLY SPRING WOULD BE THE PROUD GRANDPARENTS OF 5,598,720,000 OFFSPRING BY LATE AUTUMN. IF YOU THINK THAT'S IMPRESSIVE, WAIT . . . HE ALSO CALCULATED THAT THIS NUMBER OF FLIES WOULD CREATE A WARM BLANKET OVER THE EARTH, 47 FEET (14.33 METRES) THICK! LATER ENTOMOLOGISTS, PRESUMABLY WITH BETTER COMPUTERS, CAME TO THE CONCLUSION THAT THIS WAS A BIT STEEP, AND THAT THE 47-FOOT-DEEP LAYER OF FLIES WOULD COVER AN AREA ONLY THE SIZE OF GERMANY.

Some of these larvae will have ended up in a glass of beer parked at the end of the barbecue table and you can see them slowly sinking to the bottom, like falling autumn leaves.

Most people don't even notice these entomological phenomena. After all, the meat will be heated to kill all livestock, and the maggots that end up over the salad dressing usually crawl away to darker nooks and crannies to avoid the light. In any case, they are not known to survive the lengthy journey through *Homo sapiens*' intestinal tract.

Control

Prevention of fly strike involves keeping the animals (and their fleeces) as clean as possible: crutching, dagging and prompt treatment of wounds are some of the obvious measures, as is drenching to prevent internal parasites causing excessive soiling. Insecticidal treatments can protect the animals from maggots.

Housefly
Life cycle

MOST FOLK WILL BE aware of the fact that the housefly (*Musca domestica*; pictured below) goes through its larval cycle in a warm compost heap with plenty of decaying plant material, especially when there are some large mammalian evacuations added to the mixture. Maggots in grass clippings and dung — that sums it all up.

Around New Zealand's schools in the summer months, the smell of spilled lunches (where do all the salad sandwiches end up?) and acetic acid from kids' sweaty hands are irresistible attractions for the housefly. Acetic acid plays an important part in the maturation of eggs inside the female fly's body — they literally get high on it, and lick all the sweaty palm prints they can. Oh yes, schools and houses are the housefly's favourite places to linger. After all, where there are humans, there must also be copious amounts of waste!

FLIES AND MOSQUITOES

Benefits

As many down-to-earth gardeners know, dung and grass clippings are not really the topic of polite or erudite conversation, but without the incessant activities of our invertebrate recyclers, we would literally be up to our necks in the stuff. Moreover, mulch would be metres thick and plants would never obtain the nutrients and trace elements they need to grow and perform. So, the housefly is a very important creature in the scheme of things — always has been and always will be.

Control

Enter the La Niña winds. Flies simply hate wind. They are relatively weak flyers and always shelter out of the wind and in the sun — a bit like you and me, really. The best place for a fly to shelter is on the leeward side of a hedge. Apart from the wind shelter it often affords odd bits of nice sugary food in the form of insect-excreted honeydew. In the absence of a good hedge a fly will settle on the leeward side of your house or fence. Imagine resting on a wall with open windows and lovely household smells coming out of the dark spaces beyond — it would have to be worth investigation, don't you think?

To minimise the housefly problem indoors over summer, plant hedges as alternative shelter sites, open windows on the windward side of the house, wipe off sweaty fingerprints, and keep the house meticulously clean.

Remember the valuable work flies do in our environment — recycling is the name of the game!

HOME SWEET HOME

BEFORE HUMANS (AND THEIR EVACUATIONS) EVOLVED, THE LESSER HOUSEFLY COMPLETED ITS DEVELOPMENT IN WET BIRD DUNG, WHICH IS STILL A FAVOURED HABITAT TODAY. IN HUMAN COMMUNITIES WHERE POULTRY ARE TOLERATED AROUND HOUSES AND VILLAGES, THE LESSER HOUSEFLY TENDS TO BE A COMMON COMPANION.

Lesser housefly

WHILE I WAS VISITING the chateau at Versailles, a fly flew by. It was not an ordinary housefly, but a 'lesser' housefly — *Fannia canicularis* (pictured opposite) — the same species that lives in New Zealand. This humble fly species connects us all — a bit like the sparrow — as it lives in the general human environment worldwide, taking advantage of waste materials with a high nitrogen content. Cesspits and septic tank contents are good larval feeding sites, but the maggots can also develop in environments such as soiled nappies and cot blankets over which the odd 'accident' has been spilled. No wonder that in other parts of the world this fly species goes by the common name of 'latrine fly'.

Life cycle

Lesser housefly maggots are very nicely adapted to living in these wet (semi-aquatic) conditions. They superficially look like your archetypal maggots, but have a large number of branched, soft spines sticking out of their body. The spines, or 'processes' as they are called in entomological jargon, significantly increase the maggot's surface area and are useful as a flotation device when they mill around in wet dung — built-in water wings.

The maggotty larval forms are the feeding stages, and these are followed by pupae or puparia, which are basically the chrysalises of flies. The completion of the life cycle takes the emerging flies away from the dung heap and into the air, and that's where problems start.

In general, the flies will not go very far from where they were born. Males will find themselves a suitable site for their nuptial flights. They just love flying in the middle of a room around a hanging light fitting. The best way to describe the aerial habits of these insects is that they fly in squares, as they tend to frequently and suddenly change direction.

Fannia canicularis is smaller than the regular run-of-the-mill housefly and it is especially numerous in spring and early summer, while the ordinary housefly is a real summer-to-autumn insect around human habitation.

The species name of the lesser housefly, canicularis, is derived from the word 'canis', meaning dog, which may reflect the fact that this insect is associated with the 'dog days' of summer.

Control

If you or your neighbours have a chook run it pays to keep the droppings dry. Better still, collect the pongy, nitrogen-rich manure on a daily basis when it's fresh, and either keep it under your pillow (it clears a head-cold within 20 seconds), or chuck it into the compost bin as a natural fertiliser. At Versailles, simple pigeon-poo removal would have done the trick.

To prevent the local male flies from entering your home, install fly screens — they really are a clever investment against blundering insects and they'll allow you to enjoy as much fresh air as you like.

Those flies that do manage to get inside can often be easily killed with an

aerosol formulation. You won't need to spray a lot of product as long as you do it when the flies are active, as they take in more insecticide when flying though the toxic clouds.

Leafminer

LEAF-EATING INSECTS, SUCH as caterpillars, tend to attack their host plant by simply chewing bits off the margins or from the leaf disc. Their activities mainly take place under cover of darkness, which makes sense if you consider the number of avian predators that lurk about plants during daylight hours — it's a dangerous world if you're a caterpillar.

Sitting still during the day is a good idea and some cryptic colouration is an extra benefit. But, at some stage in evolutionary history, certain caterpillar species decided to ignore all these strategies and invent a totally new approach to protecting themselves from predators.

They started tunnelling into the leaves and eating their way through the food source, day and night — hence the name leafminers. Leafmining is a lot more common than you think and it certainly isn't restricted to the Order of caterpillars (Lepidoptera) — flies (Diptera), sawflies (Hymenoptera) and certain beetle species (Coleoptera) have all learned to mine their hosts with great success, creating arty patterns as they go (pictured below).

In New Zealand we have a number of leafminer species, most of them native types that generally explore native hosts. There are also some introduced ratbags that can cause a bit of damage to certain desirable ornamentals or food crops.

Native leafminer moths have never done any harm. Native plant hosts have grown up with their private herbivore for many millions of years. There are some rea nice ones on silver beech, olearia, karamu, lancewood and kauri.

Parectopa leucocyma, the kauri leafminer (pictured opposite) starts its mine at the tip of a kauri leaf and then heads in the general direction of the petiole, sometimes backtracking along the way. But it usually gets there in the end and not before time,

ecause near the petiole the host plant
orms a gall around the leafmining
aterpillar. This provides a wonderful,
heltered site for the larva to over-winter.
 that an example of adaptation or just
lever timing?

After hibernation, the caterpillar chews
ut of the gall and pupates on the leaf in a
archment-like cocoon, to emerge as a very
mall, but truly elegant Gracillariid moth. I
on't mind leafminers in my garden.

To be a leafminer you've got to be
mall. That almost speaks for itself, seeing
s a leafmining larva has to complete its
ntire immature life cycle in between the
pper and lower epidermis of a leaf. Often
here is not much room in that area.

It would also be handy to be a flattish
rva. Indeed, most caterpillars show dorso-
entrally flattened head capsules — flat on
he top and underside — which allows a
t more chewing-room to move.

And legs? Well, to be quite honest,
here's not much point in trying to stand on
ppy-toes inside a leaf, so most larvae have
ther very rudimentary legs or none at all.
heir movement is almost as primitive as
hat of an earthworm or a maggot.

Life cycle

The beet leafminer (*Liriomyza chenopodii*)
is a native of New Zealand and can be a
nuisance on some veggie crops — silver
beet and spinach are often the target, but
also the weed fat hen, which belongs to
the same plant group. Hence this
leafminer's very specific name.

The life cycle of these flies is fairly
standard for your average leafmining
species. The small (just over 1 millimetre
long), brown adult female flies push
their eggs into the upper epidermis of
the leaf so that the maggots that hatch
have no problem settling themselves
inside the leaf to start the long and
winding mine.

After a few weeks, and perhaps
three skin moults, the larvae are ready
to pupate. This they do in the ground,
after they have vacated their mine.

Another couple of weeks later,
the adult flies emerge from their yellow-
brown pupae, find a mate and start
looking for a suitable host to receive
the next generation of tunnelling
maggots.

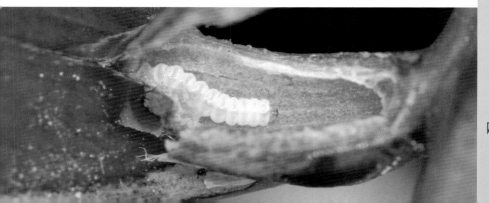

Impact

Just go out the back door and have a look at the leaves of sow thistle weed in late spring and summer, or examine the round foliage of nasturtium and you'll notice the irregular wriggly lines *inside* the leaves. These are the work of the leafminers.

A life cycle similar to the beet leafminer is completed by a related species of fly inside the leaves of nasturtium. The nice, almost-white mine gives the leaf a certain *je ne sais quoi*. On closer inspection, you can see plenty of evidence that the tiny maggot has been feeding inside — two carefully arranged rows of very dark excrement particles line the trajectory.

Our endemic kaka beak or Clianthus (red- or white-flowered — it doesn't matter) has its own special leafmining fly and it too has a species name that reflects its host (*Liriomyza clianthi*). Some gardeners get really worried when the leaves of their prized kaka beak go all white and scribbly.

The ragwort leafminer (*Phytomyza syngenesiae*) is an example of a fly species that is totally catholic in its taste. Senecio, cineraria, sow thistle and chrysanthemum are some of the common Asteraceae hosts but plantain is also on the menu. In bad infestations chrysanthemums may drop their leaves, and plants grown under glass are most at risk.

Azalea has its own caterpillar miner, a does banksia. But in the past five years or so a new species, *Dialectica scalariella* (pictured opposite), has really started to do some damage to the leaves of comfrey and that wonderful ornamental, echium.

The moth is an absolute beauty — delicate chocolate brown and white, and intricately patterned to boot. It has huge antennae — much longer than its body — and an upright stance when at rest. But the mess they cause is impressive too,

AMAZING ADAPTATIONS

ONE OF THE MOST REMARKABLE FEATURES OF INVERTEBRATES IS THEIR ADAPTATION TO JUST ABOUT ANY TYPE OF HABITAT, PLANT OR HOST. THERE'S A CREEPY-CRAWLY FOR ALMOST AN OCCASION. YOU CAN FIND THEM IN HOT SPRINGS — ALWAYS A BIT OF A TRICK IF YOU DON WANT YOUR PROTEIN TO BE DENATURED BY HIGH TEMPERATURES! THERE ARE INVERTEBRATE THAT MAKE THEMSELVES AT HOME IN THE FOLLICLES OF YOUR EYELASHES — OTHER TYPES INVADE YOUR MORTAL REMAINS SHORTLY AFTER DEATH.

THERE ARE BUGS IN BAPTISMAL FONTS, IN FORMALDEHYDE JARS IN MORGUES AND EVEN IN POOLS OF PETROLEUM. YES INDEED, THERE IS A WHOLE SAFETY ISSUE ABOUT THE CORRECT STORING AND TESTING OF AVIATION FUEL, BECAUSE TINY ORGANISMS THAT THRIVE IN THAT PARTICULAR ECOLOGICAL NICHE THREATEN TO CLOG UP THE FUEL FILTERS ABOARD AEROPLANES. SERIOUS STUFF!

specially in summer and autumn. The
eaves of their host plants are an absolute
isaster, with blotch mines between all
1e major veins.

Some species of sawflies — a plant-
eeding group of wasps — have also
earned the trick of leafmining. Their
rvae look and act a little like caterpillars.

The most visible leafmining sawfly,
Phylacteophaga froggatti, has hammered
ucalypts since it became established in
New Zealand in the 1980s. The mines are
ot nice and linear, but more like large
lotches or blisters on the eucalyptus
eaves, often with a cute, narrow 'tail'
idicating where the hatchling larva
tarted its sojourn.

It is very clever to live inside your
ost — parasites and parasitoids employ
1is technique a lot — but in some cases
1e plant hosts object to the whole idea
nd start fighting back. Galls are often a
esponse by leaves, twigs, branches or
runks to the presence of an internal
erbivore. The feeding activity of the
arasite causes the host plant to quickly
row a heck of a lot of gall material, in
vhich the offending larva is contained or
onfined. The plant literally invests in a bit

of gall growth in order to limit the
chewing damage.

Good examples of insect-induced
plant galls are the raised reddish lumps on
certain willow leaves — caused by the
willow gall sawfly — and the tiny,
numerous galls on eucalyptus foliage,
such as silver dollar gums, caused by
minute gall wasps.

But really, gall-formers are rather
crude and agricultural when compared to
the delicate techniques shown (and
intricate patterns produced) by our best
leafminers.

Control

Chemical control should really be aimed
at the larval stages, but as these are well
hidden inside the foliage you'll need to
whack on a serious, systemic compound,
which is not everybody's cup of tea.

Last year I had a go with Spinosad
(see page 14), which showed some
systemic activity on my echium
leafminer. The result was not bad at all,
but I have the feeling that you will need
to spray at least fortnightly to keep your
plants substantially free of leafminers.

Parasites

Although leafminers lead a sheltered life, concealed from most of their predators, they will not escape the attention of their long-time associated parasitoids. The eucalyptus leafmining sawfly is probably the best example of this.

A few years after the eucalyptus leafminer, a damaging forestry pest, became established its parasite arrived here too and started to do its job. The current recommendation is to facilitate the parasite by refraining from carrying out chemical control measures.

Anybody with a European oak (Quercus) or beech (Fagus) on their property has a chance to encounter the characteristic damage patterns of the oak leafminer, a Lepidopteran species that arrived here in the 1950s.

Initially host trees were quite badly affected, with many blotch mines per leaf, allowing very little room for the remaining photosynthesising cells. But natural enemies soon caught up with this exotic leafminer and the number of caterpillars per leaf is way down these days.

Mosquito

THERE ARE SIXTEEN SPECIES of mosquito established in New Zealand and twelve of these are endemic, meaning they occur only in Aotearoa. This leaves us with only four species of exotic origin, which is comfortably low and serves to remind us of the importance of our biosecurity and quarantine systems.

A lot of these species are rather attractively patterned, and if you are in possession of a low-powered stereo microscope you will not only find beautiful silver stripes on their body parts, but also scales, implanted in neat little rows along the wings' veins and trailing edge.

Some folk think that mosquitoes sting. 'Wrong end of the body,' I always say. Stinging, as practised by wasps and bees, involves a sharp rear end. Mozzies bite, and their mouthparts are quite spectacular. The term proboscis simply does not do them justice. There are six main, sharp parts: a pair of serrated knives, a pair of lancets and two very thin tubes. This series of assault weapons is kept together in a sheath and can penetrate your skin to a depth of about 1 millimetre. The serrated knives help to drive the stylets, at right angles, into the area where capillaries and small veins are most numerous. Then the stylets turn through almost $90°$ and the exploratory drilling or, more precisely, phlebotomy (a word derived from the Greek language, meaning 'the cutting of veins') continues parallel to the skin.

This exploring is carried out by thrusting and partially withdrawing the stylets many times. At the same time the mosquito injects small amounts of saliva into the area, through one of the thin tubes. The saliva contains various enzymes: some are anticoagulants, others are anaesthetics.

As soon as a number of capillaries or a small vein is ruptured, causing a small haematoma, the anticoagulant prevents the lood from clotting. This enables the mosquito to suck the viscous blood through the extremely thin tube.

If the mouthparts hit a decent-sized vein it might only take a minute for the mosquito to get a full complement of protein. A good sucker can take as much as four times her own body weight.

Stretch receptors in the abdomen notify the mosquito when it is time to pull the stylets out of the host. Consider the consequences without such receptors.

Only the females are interested in sucking blood and not all mosquito species bite humans. Some prefer warm bird blood, others go for different vertebrate hosts — human blood is a relatively new addition to the parasite scene!

It is the protein in the blood that the female mosquito needs. This protein provides the basis for egg development and most species need more than one blood meal to produce viable eggs.

The males feast on pollen and nectar to strengthen their reproductive tools. Having said that, there are some species in which the males, too, will loiter around warm-blooded hosts, simply because that's a great area to pick up receptive chicks!

Male mosquitoes tend to have larger, bushier antennae and a blunter tip to the abdomen than females. They have a clever organ at the base of their antennae that detects the vibrations of the female's wing-beat.

Life cycle

The life cycle differs slightly from species to species, but in general terms it involves water, some eggs, a few aquatic larvae, the very odd-shaped pupa and a few minute drops of blood. It starts with some water and a pregnant female.

If you take a close look at your common around-the-house species, you might be surprised at the clever tactics that these beasts employ to keep their genealogy going. A female will either lay a raft of floating eggs on a large body of water or, depending on the species, she will lay eggs just above the current water level, on vegetation or other suitable substrates that stick out of the water.

This last trick makes perfect sense. The larvae will hatch only when the eggs are inundated, so it needs to rain quite a bit for the eggs to be activated. That results in a larger pool of water, so that the survival chances of the larvae are

FLIES AND MOSQUITOES

increased — it'll take longer for the water habitat to dry up, allowing the wrigglers more time to complete their life cycle.

One native mosquito loves coastal saltwater rock pools and another peri-domestic cosmopolitan species favours dirty or slightly polluted water in gully traps and yucky drains.

When the eggs are ready to hatch, they turn into small 'wrigglers' (pictured below right) — the mosquito's larvae. They have a spiracle, or breathing opening, situated in a tubular protuberance on their last abdominal segment — that's a siphon or snorkel stuck on their bums.

All day and, I presume, all night, the wrigglers move up and down in the water. On the bottom of the pond or pool they gather organic debris on which they feed — they are the recyclers of the aquatic environment — and at the surface they stick their snorkel through the water's meniscus to gulp a breath of air.

Mozzie larvae feed in an interesting way. They stir up the water using brushes near their mouths and filter out the floating organic matter and microbes.

In a matter of days, moulting takes place. There are three larval stages in the life cycle of your average mosquito and they almost double in size with each moult.

The wrigglers then enter a pupal stage — this is the intermediary form between aquatic larva and arboreal, winged adult. The pupa looks like a fat comma that moves up and down in the water, just like the wrigglers. But the pupa, often referred to as a 'tumbler' (pictured opposite page, left), does not feed.

When the adult mosquito is about to hatch, the tumbler floats to the surface of the water. The pupal skin splits open from the back and the fresh, soft, adult mosquito will slowly emerge from its casing — the thorax first, then the abdomen, followed by the legs, one at a time.

The emerging adult swallows air, so that its body expands, which causes the pupal skin to rupture. It is an amazing sight, taking the best part of twenty minutes or so. The mosquito usually sits on the water's surface with its six legs spread out until it has hardened its cuticle (skin) and its wings before it's ready to fly.

Impact

Mosquito bites itch. They itch because of the residue of saliva that lingers under the skin. In other parts of the world this saliva

can also contain the infectious stages of some nasty diseases. Malaria and yellow fever are two of the most common and lethal diseases of the human race, killing millions of people each year. Worldwide, malaria is the leading cause of death for children under five years of age. Dengue fever, encephalitis, Japanese encephalitis and many other unpleasant diseases are also vectored by mosquitoes.

These diseases are not endemic in New Zealand because we do not have competent vector species established here.

Mosquito saliva is full of antigens and most people have some sensitivity to it — some people are more sensitive than others. Interestingly, there is good evidence that our immune system can get used to bites from local mozzies. That's why newcomers and tourists often complain about the severity of their reaction to mosquito bites.

Repellents

The research into attractants and repellents has lead to the development of a number of commercial preparations that can be used to protect humans from mosquito bites and mosquito-borne diseases.

DEET (N, N-diethyl-3-methyl-benzamide) appears to be by far the most efficacious active ingredient available but, as pointed out by concerned environmentalists, it certainly ain't an organic material. DEET has indeed been reported to have some side-effects and there are concerns about this substance.

So, what are the alternatives? Are there any mosquito repellents that are totally harmless? Some folk swear by basil extracts — the idea, apparently, came from Turkey. You could lump this under the heading of essential oils and dried tomato leaves that are in vogue with the naturopathically inclined, but generally these alternative repellents are only moderately effective for a limited time. Nothing comes close to DEET in duration and effectiveness — it's the best we've got.

Kids playing outside are seldom keen to get regular doses of repellent on their skins, but in malaria-prone regions of the world good, long-lasting repellents are important for human survival.

I'd like to add a word of warning for those of you that are intrigued by the modern devices marketed for repelling mosquitoes based on ultrasound technology — don't waste your time thinking

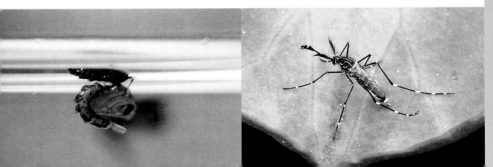

about them, because the scientific literature does not support the efficacy claims.

In New Zealand the mozzie is merely seen as a nuisance — we have the luxury of being able to be slack about personal mosquito-protection. But with the current rate of interceptions and incursions of foreign species, this may not last. A breakdown in our biosecurity could result in a catastrophe. As soon as a vector becomes established, a disease outbreak is possible at any time.

We need to increase awareness of mosquitoes. They are not just a nuisance, but a potentially serious problem. The Ministry of Health's costly actions to eradicate the southern saltmarsh mosquito (*Ochlerotatus camptorhynchus*) from the Hawke's Bay, Gisborne and Kaipara regions shows the urgency required when it comes to dealing to foreign vectors.

Control

While we may think nationally or even globally about these matters, it pays to act locally and get our own backyards under control first. The aquatic habitats for mosquito larvae are sometimes surprisingly easy to annihilate. Do you have gutters around your house? Clear away leaves to prevent water sitting in the guttering — it doesn't take much to allow a few dozen wrigglers to develop.

Moderately deep saucers under pot plants or tubs on the deck can also act as suitable reservoirs. What about barrels for collecting rainwater, or pots of lotus plants? The pool of water that sits in a child's swing made from an old tyre is a classic breeding ground for certain mozzie species. Even empty soft-drink cans do a great, dark job providing a refugium and habitat for the next generation of 'container breeders'.

Swimming pools shouldn't be a

MOZZIE ATTRACTANT

WHY DO SOME PEOPLE SEEM TO HAVE A MOSQUITO TARGET PAINTED ON THEIR BODY, WHEREAS OTHERS ARE RARELY BOTHERED BY THESE PARASITES?

FIRST OF ALL, THE CARBON DIOXIDE YOU EXHALE WILL LURE RECEPTIVE FEMALES FROM FAR AND WIDE. THEN, AT CLOSE RANGE, OTHER ATTRACTANTS TAKE OVER. SHORT-CHAIN ACIDS, EXCRETED BY THE SKIN, ARE PROBABLY THE MOST IMPORTANT CHEMICALS — STUFF LIKE LACTIC ACID — AND EVERYBODY EXCRETES SLIGHTLY DIFFERENT ACIDS IN DIFFERENT CONCENTRATIONS.

WHEN YOU ASK SCIENTISTS TO MAKE UP A MIXTURE OF THE BEST-SMELLING ATTRACTANTS FOR MOSQUITOES, IT BECOMES A LOT MORE COMPLEX THAN ANTICIPATED. IT'S DIFFICULT TO DUPLICATE THE SMELLS OF A HUMAN BODY, BUT SOME SMELL BLENDS ATTRACT, WHILE OTHERS REPEL, AND IT SEEMS THAT CERTAIN VOLATILE CHEMICALS THAT APPEAL TO MOSQUITOES AT LOW CONCENTRATIONS REPEL THE SAME SPECIES AT HIGHER CONCENTRATIONS. ONLY MOZZIES CAN TELL US APART!

roblem, as long as they are chlorinated ccording to recommended doses.

Mother Nature also provides plenty of ites — watch out for bromeliad stems, ollow bamboo, and bodies of water in he crotches of branches and in tree runks. And then there are ponds, lakes nd even water tanks, connected up to the pouting. They can all hold large quantities of water, so what are we going o do about these habitats?

Mosquitoes that develop in ponds and akes tend to prefer the shallow areas near he perimeter. Mind you, gardeners like hese shallow areas too, as a habitat for og plants and to create a very gradual ransition from a terrestrial to an aquatic abitat. The general recommendation for he prevention of mosquito breeding is to ave steep banks in ponds and lakes, so hat there are no shallow margins.

Water tanks should be made as nosquito-proof as possible, so that no emales are able to gain entry to the tank o deposit eggs. Using vegetable or nineral oils to create a very thin layer of il on top of the water has a limited effect, lthough it may block up the odd anal piracle — and what could be worse for a nozzie wriggler?

The professional mozzie busters that re going for the throat of the southern altmarsh mosquito use pellets of -methoprene, an active ingredient that nterferes with the moulting and levelopment process of the mosquito's vater-based life stages. This stuff does not mpact negatively on non-target species, asts a good few weeks in water, and can

be distributed from the air. It's a pity this control material isn't readily available to all pond owners.

Bti is another great weapon against mosquito larvae in water (see page 14).

So what about biological control? Most people do not realise that a number of fish species love to dine out on culicid larvae. A couple of goldfish in your pond will alleviate the nuisance of breeding mosquitoes. Small goldfish will even live in a lotus tub. But please don't be tempted to use that horrible mosquito-eating fish known by the scientific name of Gambusia. It is an ecological disaster waiting to happen. It is aggressive and will kill endangered native fish.

Parasitic fly
Life cycle

PARASITISM IS NOT RESTRICTED to the wasp order. Flies have been having a go at this lifestyle too, but there are distinct differences in the way in which an adult fly delivers its eggs to the prospective host. Wasps usually have an ovipositor that is strong and sharp enough to insert into a suitable target insect, delivering the egg almost like a vet would implant a subcutaneous microchip. Flies are not able to do this and therefore need a cunning plan to get their eggs inside the host.

Trigonospila brevifacies (Order Diptera,

Family Tachinidae, pictured below left) is a small but pretty striped fly that can often be seen sunning itself on foliage. It is a significant parasitoid of caterpillars, especially leafroller pests. The female can extend her soft, telescopic ovipositor forward by curling it under her body and between her legs. This way she'll be able to accurately lay a few eggs on a caterpillar, right in front of her nose!

The larvae that hatch from the eggs immediately tunnel into the caterpillar's body, where the feast begins. The empty egg shells remain stuck on the host's skin as a visual reminder of what's going on inside. They may even play a useful role in deterring other Trigonospila females from ovipositing on the same caterpillar — each caterpillar can be food for only a limited number of parasitic maggots!

Similar oviposition tricks are employed by the female of the cluster fly (*Pollenia pseudorudis;* pictured below right). This handsome blowfly-sized Dipteran, with golden hairs on the thorax, somehow manages to deposit an egg or two on earthworms close to the surface of the soil. Adult female cluster flies actively search for their host by sniffing around and entering the earthworm's tunnels.

Some of the Pales species within the fly Family Tachinidae have found another trick. They simply oviposit in such a place that the caterpillar cannot help but swallow the eggs whole. *Pales marginata* deposits up to four eggs on the edge of recent chewing damage caused by the caterpillars of the native case moth. The next night, when the caterpillar resumes feeding, it over-bites the eggs — that's an easy entry for the embryonic parasitoids. Case moths tend to be heavily parasitised by flies, especially *Pales marginata*, but also by Trigonospila.

Parasites

Mind you, some parasitoids are often parasitised themselves. A good example of such a secondary parasite (or hyperparasite) is the Eupteromalus wasp that gets into *Pales marginata*. How the hyperparasite finds its elusive target and then manages to lay a few eggs inside the maggot, which feeds inside the case moth caterpillar, which sits inside the sturdy silken case, is still a mystery to me. But the telltale signs of the hyperparasite can be easily found: look for the perfectly round emergence holes in the side of the cases.

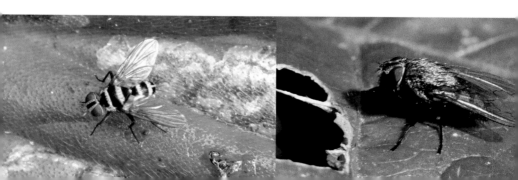

Slimy molluscs

GASTROPODA IS THE NAME for the Molluscan class of animals that includes slugs and snails. The name Gastropoda means 'gut on a foot' and I suppose most growers will agree that it is a well-deserved label for these damaging critters.

The diet of slugs (pictured below) and snails (pictured above) can be varied; some specialise in certain groups of plants, others are general scavengers and composters, whereas some native species are downright predators. The most damaging groups of gastropods are the introduced slugs and the brown garden snail. Somehow these creatures seem to sniff out all the tender plants in the veggie garden.

There's one factor that rules the lives of slugs and snails — moisture or, rather, the lack thereof in dry periods. Gastropods that live in water do not have the same problems as our terrestrial friends. Drying out is a real risk on land and snails have adopted a great technique for avoiding this potentially lethal condition.

During dry periods slugs, advantaged by their sleek, shell-less body, can seek moist sites in the top layers of soil. That's probably why they still perform some damage in the middle of a dry summer, on kumara, carrots and potatoes. Living underground in a moist environment that's also a food source . . . clever stuff!

Symmetry has never been a gastropod's strongest characteristic. Snails' shells are usually asymmetric and even the bodies of slugs have unbalanced design features. The breathing hole (also known as the pneumostome) tends to be off-centre, on the right-hand side of the body. It opens and closes at regular intervals, albeit in slow motion.

Speed is another thing that slugs and

snails are very unfamiliar with. They move by producing copious amounts of slime or mucus, on which the large head-foot slides forward. It takes a wee bit of time to produce that mucus and it also depletes the moisture levels inside the body, so there simply isn't a lot to go hooning with.

During daylight hours I once clocked a garden snail travelling at about 12 centimetres per minute — certainly not enough to set off the speed camera — but on wet, dewy nights the moisture in the environment will allow speed and foraging range to dramatically increase.

Of the 35,000 species of terrestrial gastropods on our planet, New Zealand has an impressive 1350. Most of them are native and they are all clever at finding food, plus pretty astute at coping with dry conditions and hermaphroditic lifestyles.

A mere 29 species (15 snails and 14 slugs) are exotic, having successfully established themselves here, and it is these exotics that cause the main problems in crops and ornamental plants.

When cruising undisturbed, you can see the 'foot' doing its thing. But if you really stop to think for a moment, you'll soon realise that this foot is merged with the mollusc's head. Scientists imaginatively refer to it as the head-foot.

Slugs

THE PHRASE 'A SLUG is merely a snail without a shell' is not far off the mark at all, but looks can be deceiving. There are many species of slugs that still have a shell of some description, although it may be reduced to such an extent that it is unable to contain the body of the owner. Some slugs have a hard shell — a bit like a fingernail — on the top of their mantle. Others have a rudimentary shell, or fragments thereof, inside their body.

Impact

This chewer is easy to identify. It creates holes in leaves, and leaves behind frass and a distinct slime trail.

Control

Remove obvious shelter sites and damp places where dehydration-prone molluscs

an hang out. It's also a good idea to
remove rank grass and patches of tall weeds
from the gastropod-threatened growing
areas. Bare soil has a tendency to be a lot
drier than covered soil, and dry materials are
generally avoided by slugs simply because
they have to produce a lot more mucus
and hence lose a lot more moisture) to
traverse those substrates.

The use of baits can be tricky,
especially when dogs are around. Make no
mistake, when ingested these baits act very
quickly and are often fatal. Believe me —
I've made the dash to the vet!

Ideally, baits should be delivered in
tamper-proof bait stations, so that non-
target species have no hope of getting
access to the pellets. Baited take-away
bottles (pictured opposite left) with cut-
out holes secured with bamboo stakes (and
perhaps even covered by some rocks or
heavy bricks) are ideal and will certainly
attract the gastropods. Here they'll die, out
of reach of warm-blooded predators.

I have found beer traps to be rather
inefficient for slug and snail control, which
is a pity, seeing as you could make a lot of
good jokes about gassed pistropods!

Predators

I'm not too sure if encouraging thrushes and
hedgehogs into your garden patch is going
to make significant dents in the snail or slug
population, but you can have a go at it.

Lure the hedgehogs into your garden
with diluted milk and bits of mince meat
(but watch for klepto-competition from
tiger slugs).

Tiger slug

THE INTRODUCED TIGER SLUG
(*Limax maximus;* pictured opposite right) is
native to eastern and southern Europe and
has been in New Zealand since the 1870s.

This large slug is generally attracted
to decaying organic material and fungi
but it certainly frequents compost bins
where kitchen scraps are on the menu.
As a good, all-round recycler, dead
animals are also part of its diet.

Can you guess where this is heading?
Ever stepped on a large, slimy slug at 3am
in the middle of the kitchen floor, groping
for a piece of cheese from the fridge?

Tiger slugs will cruise your lino at
night, attracted by the remnants (or
minute spillages) of your cat's meat.
They've even been known to slurp the
cat's milk. After all, protein is protein,
and 'recycling' is Limax's middle name.

But, unless you hate stepping on
naked gastropods in the middle of the
night, the tiger slug is really not a pest at
all. What's more, they have a fascinating
nuptial behaviour. When ready to mate,
tiger slugs find each other at night, circle
for a while and frequently touch bodies
and produce large amounts of mucus.
This mucus is so strong that it can be
stretched into a rope, from which the
intertwined slugs dangle for 15 minutes or
more, while exchanging other packages of
body fluids. Tiger slugs dangle from
branches, ledges or decks — who knows,
the Auckland Harbour Bridge might be
next: mating while hanging from a bungy
cord!

SLIMY MOLLUSCS

Snails

THE SNAIL IS FAMOUS for having four tentacles protruding from the 'head' region of its head-foot. Two are ocular peduncles, or in normal language, eyes on stalks (pictured below right). These peduncles can be retracted — inside-out — when the snail perceives danger. Sounds a little bit like an ostrich trick to me, but since this invertebrate usually also facilitates a total withdrawal of its soft body parts into its hard shell, it must be a successful protective measure.

Everybody wants to know where the snail's teeth are. Well, those teeth are something quite special; there can be as many as a few thousand so-called radular teeth on the radula (the snail's tongue). This tongue can be extruded from the mouth, which is situated on the sole of the head-foot. All those hard teeth are used to unceremoniously scrape and erode the soft plant materials on which the mollusc feeds.

When teeth are worn, new ones are produced from the back of the radula.

This replacement is continuous and ensures the uninterrupted ability to feed.

Life cycle

Whereas a lot of invertebrate animals practise the art of hibernation, sitting out winters in a state of torpor, gastropods turn things around and aestivate. During the dry days of summer snails can often be found in cavities, such as the in-ground water meter, or on the ceiling of the letter box (with semi-destroyed junk mail as a dead give-away of a gastropod's presence).

As the weather becomes warm and dry, snails retreat into their shells and construct thick mucus layers just inside the opening of the shell. The mucus hardens to be like a plate-glass window and acts as the perfect membrane to stop dehydration while still allowing air to permeate. The moment relative humidity increases (due to long periods of rain, for instance), the mucous membranes dissolve and the snails wake up from their aestivatory sleep.

Courtship in gastropods is fascinating. While they are hermaphrodites, carrying both male and female genitalia, they can't

do it by themselves. You need two specimens that line up, more or less back to front (pictured below).

Garden snails circle each other during foreplay, both partners grazing, biting or nuzzling each other. At the same time they lay down a platform of thick mucus. Next they start a new Olympic sport — dart shooting. Calcareous darts are made from glands in the garden snails' bodies. The aim is to penetrate the partner's head-foot with those darts . . . just for a bit of stimulation, or so scientists think!

The exchange of spermatophores (little packages of sperm, wrapped in biodegradable Glad Wrap) finalises the deal and both partners split, leaving a messy, slippery site for the gardener to stumble upon in the cool of a spring morning.

In winter and early spring a lot of people find fresh slug and snail eggs (slug eggs pictured opposite left) all around their crops at soil level. They're clear and translucent and surprisingly large. Other telltale signs are dense patches of silvery mucus in one spot, or slithers of slimy strands hanging from branches.

Impact

What, apart from junk mail in letter boxes and wine labels, does your average garden snail feed on? It appears they are rather selective in their preferred food plants; as a general rule of thumb a succulent little number will always be preferred over a rough-and-tumble type of plant.

They also go for dead plant tissue and even algal and fungal growth. Climbing shrubs are not a problem, especially if there are new leaves to be had. Citrus is also a surprisingly good host for *Cantareus aspersus*, the brown garden snail.

Control

As with slugs, remove obvious shelter sites such as rank grass and patches of tall weeds and eliminate damp places.

Baits should be delivered in tamper-proof bait stations.

What about sharp sand or, better still, sharp eggshell fragments around the threatened plants? These snails, with their

soft head-feet, surely won't be able to cross such a viciously sharp substrate?

To answer this question I have done a wee experiment. I let a snail cross the ultra-sharp blade of a brand-new craft knife and observed its fate (pictured below). Of course, the sole of the snail's foot is never in contact with the substrate — there's always a layer of mucus in between.

Besides, what do you think eggshell is made of: calcium carbonate, or lime. 'Hmmmm!' says our brown garden snail. 'Now, that's what I call service . . .'

Predators

Again, you can try to encourage thrushes and hedgehogs into your garden patch.

Provide the thrush with a few 'anvil' stones on which it can enthusiastically smash the snail's hard, calcareous shells to pieces. Lure the hedgehogs into your garden with diluted milk and bits of minced meat. But be aware that 'encouraging' the introduced hedgehog is increasingly seen as an act of eco-ignorance or even eco-vandalism.

Brown garden snail

CANTAREUS ASPERSUS, THE MOST common garden snail, was introduced to New Zealand by the French in the early 1860s. The reason for this is really simple: the brown garden snail is a highly prized food in the Mediterranean region, where it is known as *escargot*. In fact, it has always been the preferred back-up for the slightly larger, but closely related, vineyard snail (*Helix pomatia*), the number one gastropod delicacy in France.

Orchid snail

A RELATIVE NEWCOMER IS the orchid snail, *Zonitoides arboreus*, arriving in the middle of last century. People who love orchids will know this little beggar well, since those plants and tender ferns form the bulk of its diet.

Zonitoides loves the warm, protected habitats provided by greenhouses, although it thrives in the outdoor environment in

warmer climates, especially from Whangarei north. According to some growers, it can be numerous and hard to control. Their small size allows them to hide in the smallest nooks and crannies and traditional baits are not very effective. Good greenhouse hygiene may be the answer and perhaps the odd treatment with Mesurol spray.

Kauri snail

THE KAURI SNAIL (PARYPHANTA species), although restricted and patchily distributed, is quite at home in the northern North Island. It feeds on earthworms by sucking them out of the ground like spaghetti and enveloping them with the head-foot. The radular teeth do the rest.

Giant land snail

OTHER NATIVE GIANT LAND snails (Powelliphanta species; pictured opposite right) dine on a similar diet of live animals, including slugs, snails and other soft-bodied invertebrates. These beautiful snails are very rare indeed — and rodents, pigs and hedgehogs regarding them as tasty *escargots* doesn't help their population. These magnificent giants can reach considerable ages (a couple of decades at least) and appear to have a relatively low reproductive rate.

EATING SNAILS

IF EATING SNAILS IS YOUR THING, THEN COLLECT THE FINEST, FITTEST AND FATTEST SNAILS FROM YOUR GARDEN. PUT THEM IN A BIG JAR AND FEED THEM FOR FOUR TO FIVE DAYS ON OLD WHITE BREAD. THIS FEEDING IS AN IMPORTANT PROCEDURE TO EVACUATE THE SNAILS' GUTS BEFORE COOKING. THE WHITE BREAD SLOWLY REPLACES THE USUAL DARK EXCREMENT, IMPROVING THE TASTE OF THE ESCARGOT BEYOND BELIEF!

NEXT, PUT A BIG POT OF WATER ON TO BOIL AND CHUCK ALL THESE CLEANSED SNAILS INTO THE BOILING WATER AND SIMMER FOR ABOUT SIX MINUTES. DON'T OVERCOOK THE SNAILS OR THEY WILL HAVE THE TEXTURE OF RUBBER BANDS.

TAKE THE SNAILS OUT OF THE POT AND LEAVE THEM TO COOL BEFORE REMOVING THE BODIES FROM THEIR LONG-TERM RESIDENCES WITH TWEEZERS.

NEXT, FRY THE SNAILS IN SOME GARLIC BUTTER FOR FIVE TO SIX MINUTES. AGAIN, DO NOT FRY TOO LONG OR THEY WILL BE LIKE GARLIC-FLAVOURED RUBBER BANDS. YOU CAN ALSO BED THE SNAILS ON GARLIC BREAD SO THAT THEY ARE MARINATED ALIVE!

SERVE THE SNAILS IN THEIR OWN SHELLS. I CAN RECOMMEND SERVING THEM WITH A 1965 CHATEAU TAHBILK — THEY GO WELL TOGETHER. BON APPETIT!

SLIMY MOLLUSCS

Sucking bugs

To get an impression of the modus operandi of sucking creatures we need to look at how a vascular plant operates.

Water and nutrients are taken up by the plant's roots and transported through the wood via xylem to the green leaves. Leaves have the ability to take up carbon dioxide and they utilise the sun's rays to provide energy for the most amazing chemical process on earth: photosynthesis. This process combines the raw nutrients with the carbon dioxide and produces oxygen and carbohydrates (sugars), which form the building blocks of new plant growth: new leaves, flowers, fruits etc.

The sugars are transported throughout the plant in vascular bundles called phloem, which pass through trunks, branches, roots, twigs, petioles and the leaves' veins, so the precious sweet cargo can be found right throughout the plant in just about all its tissues. And that's quite handy if you are a sucker. All you need to do is use your sharp, tubular mouthparts to hit one of those rich phloem veins . . . then sit back and enjoy the sweet viscous liquid that will literally flood into your mouth.

It's not quite that simple, but you get the idea: a sucking insect taps into the phloem sap to extract the nitrogen and certain proteins it needs for its own development. To do this more efficiently, a sucker often increases the phloem pressure by injecting some saliva into the plant.

If we think of sucking insects as walking or flying hypodermic needles, it suddenly becomes very clear what the consequences could be for the host plant's

ealth: the combination of hypodermic
eedles and veins can also lead to the
ansfer of nasty diseases, especially viruses.
iruses may not always be deadly but,
ore often than not, they can severely and
ermanently debilitate a plant or its yield.

Aphid

OME APHIDS HAVE FOUR very feeble
vings, which makes you wonder how on
arth these small and fragile insects are
apable of sustained flight. Winged aphids
isperse as adults in search of a new host
lant, but even on a still, warm and sunny
fternoon aphids won't fly very far —
erhaps a few dozen metres — so it pays
) keep an eye on your garden and crops.

The mere fact that aphids develop
vings is a strange story in itself, for when
ou look at an aphid colony on your
rized plants you'll notice that most of
em have no wings at all. Only on very
eavily colonised plants, where aphid
opulations build up in tremendous
umbers, do the larvae start to develop
heir little wing-buds. This is caused by a
ombination of a few events.

When a lot of aphids tap into the
same sap stream of the phloem, the
quality of their tucker diminishes. When
the population densities become so high
that the poor larvae frequently 'rub
shoulders' with their siblings, it is time to
colonise new hosts. These are two pretty
good reasons to grow wings and take off,
and you can not blame them for using
the powers of incomplete metamorphosis
to further their expansionist behaviour.

If you think an aphid is small, you
should see its proboscis. It's a specialised
sucking tube made from carefully
arranged mandibles, and consists of two
parallel tubes. Through one of the tubes
the insect sucks up the viscous sweet
phloem sap, through the other it injects
saliva into the plant. The saliva secretions
help to break down starches and cell
walls, which makes sucking a lot easier.

Life cycle

If you look really carefully — perhaps with
the aid of a hand lens — at a colony of
aphids it becomes clear that the population
consists of some large aphids (some of
which may have wings) and a battalion of

SUCKING BUGS

smaller specimens, often positioned in nice 'undulating' groups, almost as if they are arranged by age and size. And that's exactly the way it is.

The big specimens in a colony are the mothers of all the baby daughters. Every day a mature female aphid produces up to five more aphids. The curious thing about this is that these mothers do not lay eggs, but live young. The egg stage has been reduced to such an extent that they hatch while still inside the female's abdomen. This enables aphids to quickly colonise new host plants. And a few live babies per female per day adds up rather quickly and alarmingly to an infestation.

The female nymphs that are born have embryos inside their bodies already, and it takes about a week to ten days before they are ready to reproduce. Exponential growth may be a cliché in the investment industry, but I think it was really the aphids who invented the concept.

To speed up the colonising process even more, these female aphids have adopted another strategy called parthenogenesis. Those of you with some classical languages under your belt will recognise this word to mean something like 'virgin birth', and this is probably the simplest way to describe the process. What process? There's hardly any 'process' left, and it really worries me (as a boy entomologist) that these female aphids are able to reproduce without the generally attentive services of the male aphids. What frightens me even more is that partheno-genesis has become a real trendy pastime in some other entomological groups as well (some species of weevil, stick insect, mealybug, and scale insect also practise th shortcut).

Impact

The visible consequences of an aphid attack are related to the interception of th

hloem sap. If a whole army of these
uckers remove the plant's carbohydrate
ood, the result upstream can be a bit
isastrous. The new growth gets the look
f deficiency. Yellowing, stunting and
ven curling or distortion of the new
aves and flowers are often part and parcel
f the symptoms caused by vandalising
phids. Remember, these insects intercept
ie nitrogen from the sap flow! And there
ill be sooty mould, the matt-black
ingus that feeds on honeydew, a clear
veaway that there is a sucker present.

While aphids benefit from mainlining
ι infected plants it come as no surprise
ιat aphids and viruses have a pretty good
lationship, built over a few decades of
:ological time. Aphids are excellent
ectors of viruses. First of all, there's the
liva injection into the plant. Certain
ruses are able to live inside the aphid's
ιdy and end up in the salivary gland,
ady to be spewed out into a new, clean
ost. Secondly, aphids can be 'passive'
ectors, simply by having a dirty or
ontaminated proboscis. Aphids actually
enefit from this habit, as virus-infected
ants produce relatively more nitrogen in
eir phloem sap.

Control

ontrol can be achieved by spraying a
hole range of registered insecticides;
ısically, any chemical that hits them as a
ontact insecticide will kill these soft-
ɔdied creatures. However, systemic
secticides (those that are taken up by the

plant and transported through the
vascular system) tend to work the best.
It always pays to read the label claims
of insecticides before choosing your
weapon. Some systemics, such as Gaucho
and Chess, work as an anti-feedant.
If you want to get a bit more organic,
simple spraying oil will do damage,
as does the old favourite, pyrethrum.

It is interesting how a lot of people
are still fascinated by the old aphid
remedy, no doubt handed down for
generations, involving the use of the soap
suds from the dishes or washing machine.

APHID CONTROL

PURE PARTHENOGENESIS IN APHIDS CAN
HAVE A FEW DRAWBACKS: THERE ARE
NO SUITABLE OVER-WINTERING STAGES
(EGGS OR PUPAE) IN THE LIFE CYCLE AND
THE ABSENCE OF MATING COULD LEAD
TO GENETIC DEGENERATION AS THE GIRLS
PRODUCE CLONES OF THEMSELVES. THE
CUNNING APHID HAS A WONDERFUL
SOLUTION TO THIS POTENTIAL PROBLEM,
HOWEVER: IN AUTUMN, WHEN THE
WARM SEASON COMES TO AN END,
APHIDS SPONTANEOUSLY PRODUCE MALE
OFFSPRING. THESE WILL MATURE AND
MATE WITH FEMALES. THOSE MATED
FEMALES WILL THEN LAY EGGS — JUST
IN TIME FOR WINTER. HOW CLEVER IS
THAT? VOLUNTARY PARTHENOGENESIS —
ONLY PRODUCING MALES WHEN
YOU REALLY NEED THEM!

The problem is that these days the detergents are not what they used to be: they clean the aphids rather than kill them! So, when it comes to using these kinds of products it pays to go back to the old formula of hard, abrasive soaps, or else obtain the modern reinvention of them — the fatty acids or potassium salts, also known as insecticidal soaps. Nice stuff: it's abrasive (causing skin damage, leading to water loss and dehydration) and particularly nasty to small, soft-bodied pests, but generally not too detrimental to the hard-bodied predators and parasites.

It is important to realise that most insecticides are not very selective; they will harm beneficial insects as well. So supreme vigilance and surveillance, followed by digital control will save you a lot of hassles — get Mama before she even thinks of parthenogenesis.

It has been well established that a number of small invertebrates are somehow attracted to the colour yellow. The most famous example is perhaps the whitefly in greenhouses, but aphids are also in that category. Use yellow containers (painted ice-cream containers or photographic developing dishes will do), filled with a mild solution of dishwashing liquid to lure and drown adult aphids.

These days sticky yellow traps are readily available in garden centres and the sticky stuff can be bought in aerosol cans for ease of application. Alternatively, you can use grease and other tacky materials t complete your lethal device — your catc will be surprising!

Predators, parasites and pathogens

There are, thankfully, a whole range of natural enemies for the aphid. Busy little silvereyes will systematically search for these small suckers, probably lured by the deposits of the equally attractive honeydew. No doubt other small insectivorous birds will also consume the odd kilogram of the little vandals.

Invertebrate predators include juveni praying mantids, which will cruise up a stem and 'graze' the aphids as they go. Ladybirds are also capable of destroying

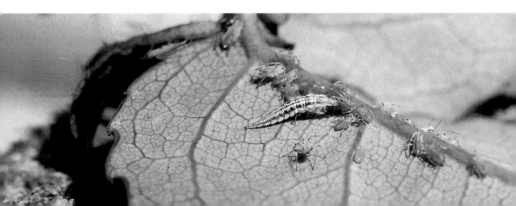

ozens of specimens each, but it must be noted that it is especially the larvae of these ladybirds that consume the most. Adults will dine on them too, but to be quite honest, they are really more interested in finding a partner. It is therefore of little use to import orange (eleven-spotted) ladybirds into your rose garden, as all they'll do is spread their wings and fly off in search of a mate.

The larval form of the Tasmanian lacewing (*Micromus tasmaniae*; pictured opposite) is a useful predator. It has long, calliper-like mandibles that can pick up and perforate an aphid with amazing speed and agility.

The maggots of some hoverflies slither through the colonies, causing mortal mayhem amongst the flock. The maggots' mother — a pretty hoverfly species (pictured below right) — will lay her eggs in the vicinity of the food source, so that her kids have only to stretch out for their first snack of aphid burger.

In the parasite department we are helped by the activities of a range of species of tiny wasps. It is truly a small miracle (with the emphasis on the word

small) that, after mating, the female wasp manages to find a half-grown aphid and parasitise it by carefully laying one egg inside the bug's body. The egg hatches inside the aphid and becomes a small wasp larva. This larva will feed on the host's internal body parts and complete its life cycle within a few weeks.

When the parasitic wasp pupa hatches, it chews its way out of the swollen and 'mummified' aphid skin, leaving a neat little exit hole. Dead aphids with little emergence holes in their bloated bottoms are a good telltale sign of parasite activity in your aphid colony.

In some cases you'll find that aphids appear to have been attacked by a fungus. The cabbage aphid seems to be especially prone to infections by *Entomophthora aphidis*. It is likely, however, that this fungus (and other species of entomopathogens) becomes more noticeable when there are high population densities of aphids on the plants, or when the plants are grown in glasshouses where a generally higher relative humidity assists the efficacy of spore germination.

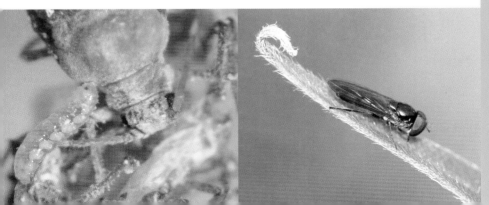

SUCKING BUGS

Cicada

IMAGINE HAVING TO LIVE for years and years in a dark, subterranean habitat, clambering through bits of clay and stones, searching for your next meal in the form of a juicy root. Imagine having to construct an underground cell at each location, and all you've got to work with are your forearms. When you finally emerge from the soil, wouldn't you be happy for a bit of R and R with your mates?

Each summer, right through to autumn, you can hear cicadas having fun, especially in areas with plenty of trees. The noise can be deafening when male cicadas are looking for a partner; some of the larger New Zealand native species get well over the 90-decibel noise level.

The noise cicadas make differs from species to species. Even closely related varieties can be separated simply by making a recording of their song and running it through a sonograph.

When you examine a male specimen and look on the underside of its abdomen you find two fairly large, round 'tympani', which act as amplification modules for the sound produced by rapidly contracting muscles. Underneath these tympani there is a ribbed membrane. Cicadas can vary the noise they make by alternating sound from their left and right sides. It is as if it is playing with a graphic equaliser.

Some New Zealand cicada species augment their love song by rhythmically 'clapping' their wings against their body or the branch they are perched on. If you get the chance, observe a clapping cicada such as *Amphipsalta cingulata* in its seemingly endless concert — it's fascinating stuff!

Life cycle

Adult cicadas invariably do the adult thing resulting in gravid females searching for a good site to lay eggs. Unfortunately, these oviposition sites tend to be pencil-thin twigs of generally desirable or economically important shrubs or trees. In late summer and autumn one of the regular and predictable questions asked of entomologists is in relation to cicada egg-laying scars (pictured below).

The females are equipped with a sharp, sword-like ovipositor, which is thrust into the bark and wood of the chosen twig. Once one egg is laid, the female moves one short step backwards and thrusts another egg into the twig, next to the first one. After a dozen or so eggs are laid, she'll fly off to find another suitable site, but in the meantime, the damage is done.

Laying eggs inside a living branch has its advantages. First of all the eggs are protected from predators, although small parasitic wasps seem to be perfectly able to

ind them. Secondly, it prevents them from drying out before they hatch — living wood is always moist. How long it akes for the eggs to hatch varies from pecies to species, but it is not unreasonable to expect periods from a few veeks (which would mean an autumn hatch) to six months (spring), or even a ear (summer–autumn).

The white to pale cream first instar arvae, also known as cicada nymphs (pictured below), drop from the twig in vhich they were born and end up on the oil, which they enter via cracks and revices. They're able to dig themselves leeper and deeper with the aid of specially dapted tibiae on their forelegs. These are ike large bolt-cutter devices with powerful nuscles and sharp edges, reminiscent of the gadgets worn by Edward Scissorhands in he film of the same name.

Cicada nymphs may be almost blind nd appear odd and fragile, but they do have the tools for the job. As soon as they tumble across plant roots in the soil, they ut them and insert their stylet, or sucking ube, into the vascular system.

The nymph's stylet comprises two adjacent channels: one for pumping saliva nto the root to create a bit of pressure, which allows plant material to be sucked up through the other.

While thrashing around the roots, nymphs create a cosy, smooth, earthen cell for a temporary home, with plenty of room to turn around in. All this is hard work and the rewards are minimal in terms of quality nourishment. Small wonder that the life cycle of the cicada takes a long, long time to complete.

The small species spend at least two or three years as nymphs underground, while the larger chorus and clapping cicadas will take as long as seven years to complete their larval stage. They may moult five times, growing in size with each change of skin. This explains why gardeners can stumble upon nymphs at any time of the year when they dig into the perennial border.

The emergence of the pre-adult nymph from the soil is not often witnessed, as it mainly happens at night. The insect resembles a large version of earlier instar nymphs, complete with bolt-cutter forelegs, but it is fawn brown, has bulging, well-developed eyes, and shows distinct wing-buds over the dorsal abdominal segments. These wing-buds contain the shrunken rudimentary

SUCKING BUGS

INSECT LARVAE HAVE A WEE BIT OF A BUILT-IN PROBLEM: BEING INVERTEBRATES, THEY DO NOT POSSESS AN INTERNAL SKELETON. IN ORDER TO KEEP THEIR SHAPE — AND PREVENT A TOTAL COLLAPSE LIKE A LIMP BEANBAG — INSECTS HAVE DEVELOPED AN EXOSKELETON. THEIR SKIN ACTS LIKE A SKELETON.

THIS SKIN CAN BE SUPPLE, OR HARD AND SCLEROTISED — IT MAY BE ABLE TO STRETCH A BIT WHEN THE ANIMAL FEEDS ITSELF SILLY AND 'PUTS ON THE BEEF', BUT THIS SKIN CANNOT GROW. THAT IS A REAL BOTHER WHEN YOU ARE A MAJOR EATING MACHINE, BUILT FOR GROWTH. IN ORDER TO ACCOMMODATE ALL THIS GROWTH, THE LARVA SHEDS ITS SKIN FROM TIME TO TIME AND MAKES A NEW, BIGGER SKIN FOR THE NEXT 'INSTAR'. SOME SPECIES MOULT THREE TIMES AS LARVAE; OTHERS CAN DO IT TWENTY TIMES OR MORE.

Impact

The oviposition scar made by the aptly named chorus cicadas (*Amphipsalta zelandica*) is unmistakable: a long, narrow series of cuts, roughly in the shape of a herringbone pattern. This scar significantly weakens the twig or branch, often causing a break during periods of strong winds. This may not be a big deal when it occurs on native shrub or tree hosts, but when a fruit tree is subjected to this form of uncontrolled pruning the losses in terms of yield can be measurable in years.

Control

What can we do to stop those cicadas laying their eggs in my valuable trees? The answer is simple: nothing. Or, if you're doing radio talkback: 'Nothing, unless you can predict exactly where the female is going to lay her eggs, because in that case your options are literally endless . . . slingshot, air rifle, local application — by paint brush — of some Napalm, etc.'

Predators

Any insect that draws attention to itself with constant noise is a dead sitter for an inquisitive predator. Often you can see a sparrow, starling or mynah flying off with a frantic cicada in its beak — the sound track literally streaks across the sky! On many occasions the chirping insect manages to free itself from the grasp of the predator, and I wonder if that happens because the bird gets a numb beak after a few minutes

versions of the adult cicada's wings.

The nymphs walk up a stem or trunk of a tree and grab a good foothold. Then the larval skin splits open via a suture on the back. The adult inside literally pumps up its wings to open the skin even further, before it can emerge. The finishing touches, in terms of hardening of the skin and final wing inflation, are usually completed before the sun comes up in the morning.

of exposure to the cicada.

Life in the soil can create a few problems for the slow and cumbersome nymphs. Predatory beetles and their larvae are always a danger to one's health, as are various fungal diseases. Beauveria is the fungus that causes cicadas to turn fuzzy-white, usually killing the insect in the adult stage, although the infection takes place inside the nymph underground.

Surprisingly, drowning in saturated soil does not appear to be a common occurrence as nymphs can survive a flooded cell or burrow for something like four weeks!

When larval growth is nearing completion, the last instar nymph slowly moves to the upper layers of the soil in preparation for emergence, where they become the most important dietary item for our endangered kiwi.

beetles but they sure do stink. Their unpleasant odour, excreted from abdominal glands, plays an important role in turning off predators.

The smell also acts as a communi-cation device that signals alarm when bugs are disturbed by a predator or other threatening organism — it relays the message of panic, causing the bugs to drop to the ground and 'play possum'.

These exotic suckers were first found in New Plymouth in 1944 and quickly spread throughout the warmer parts of the country.

Life cycle

In winter you find green vegetable bugs in amongst the terminally dead vines of courgette plants and under the crisp debris of the previous summer's scarlet runner beans. The insects move rather

Green vegetable bug

NEZARA VIRIDULA, THE INTRODUCED green vegetable bug, has three characteristic light dots on the front edge of the triangular shield in the centre of its back.

Despite the fact that many New Zealanders call these insects 'green shield beetles' or 'stink beetles', they are not beetles at all, but true bugs of the Order Heteroptera. And they might not be

SUCKING BUGS

slowly and show none of their warm-weather, jumpy habits. They also sport a different colour. The bright green body changes to a dull, dark brown-green to purple as the bug takes time out in winter. Life slows down almost to a standstill and feeding ceases.

When spring temperatures rise, the successfully over-wintered adults kick-start their feeding and look for a mate. This results in the characteristic batches of eggs that provide the first generation of larvae or nymphs in spring.

The eggs are unmistakable clusters of 50 to 80, laid in neat little rows on the underside of leaves. Initially the eggs are a pale yellow, but they darken as the embryo develops over the next fortnight or so.

When these eggs hatch the smart, dark, first instar nymphs waste no time sticking their sharp stylets into the phloem bundles of suitable host plants, the list of which is enormous.

Many people are confused about those small, circular and predominantly dark brown to black immature insects they see on their plants (pictured below). Indeed, it is not obvious that they are the nymphs or immatures of that common, green, smelly stinkbug. And the fact that these nymphs change colours and spot patterns with every skin moult does not help matters.

At first the general colour is black with rows of white spots on the abdomen. Later instars become more colourful, with red marginal spots and creamy-yellow spots on the middle parts of the abdominal

segments. At the same time, the general colour of the abdomen gets a green tinge, and the black wing-buds grow with each change of skin. Just before the nymphs turn into fully winged adults, they are green with red legs, sporting a multitude of colourful splashes along the margins and in the centre of their bodies.

Impact

During the summer months, symptoms of infestations range from the shrivelling and distortion of developing seed pods (beans and macadamia nuts) and fruit, to the discoloration and wilting of plant stems. These green vegetable bugs literally suck the proverbial out of plant material.

Beans, tomatoes and other solanaceous plants, brassicas, cucurbits, spinach and silverbeet are some of the commonly attacked crops, as are passionfruit, tamarillo, grapes and a wide range of ornamentals.

Control

Digital control is the way to go — it's cheap, easy, and uses the insect's own communication device. When you start squashing green vegetable bugs, you'll notice the effect of the released warning signals immediately. Keep squashing whatever bugs fall on the ground. This way the concentration of the smell increases, followed by the number of 'jumpers'.

Homeopathic insect controllers advocate making a brew of Nezara's own juices. It's worth a try!

Predators and parasites

Traditionally, gardeners would chuck all sorts of systemic insecticides at these pests, but I prefer to be more cautious and use Nature's natural controllers: the predators and parasitoids.

Trissolcus basalis is a tiny wasp that was introduced from Egypt in 1948 in an attempt to control the green vegetable bug. The female Trissolcus lays one egg in each of the bug eggs, and often the whole cluster will be parasitised.

It truly is a microscopic affair: the whole life cycle of the parasite takes place inside the egg and there's room for only one developing wasp.

Unfortunately, *T. basalis* does not only parasitise Nezara eggs, but will have a go at various related Pentatomid species, including the predatory soldier bugs. In any case, the levels of parasitism are by no means high enough to make a significant dent in the population of the green vegetable bug.

When you are working away in the garden in winter or early spring, and you find these poor, struggling stink bugs, trying to get through the winter alive, you know that each one of these is potentially capable of producing dozens and dozens of offspring on your vegetables or fruit.

One quick, sharp pinch of the thumb and forefinger is all it takes to prevent that from happening and bring the spring population down! Now if that isn't economic digital control . . .

SUCKING BUGS

Mealybug

MEALYBUGS LOVE TO LIVE in the nooks and crannies of a plant, places where they can hide from the public eye and go about their sucking business without disturbance. Look on the underside of leaves if you suspect these insects are diminishing the vigour of your prize botanical specimens; the white waxy meal will soon give the game away.

They are usually mobile and tend to be white. Their bodies might be brown or pinkish, but the waxy secretions that give them their protection is usually pure white (pictured below). Some have long, showy tail filaments that trail behind the insect from a very early age (long-tailed mealybug; pictured opposite).

Life cycle

Mealybugs are most numerous in the warmer periods, especially late summer and autumn. Some species practise parthenogenesis. Eggs or, in some instances, live nymphs are deposited in waxy or mealy 'nests' on the host plant. Many generations can be produced in one season.

Control

If you've ever had to contend with honeydew on your carpet beneath a mealybug-infested hoya vine or potted palm, you'll know that you've got to get rid of them . . . fast! But waterproof sucking insects are a bugger to control; water-based insecticides simply run off them like water off a duck's back.

Non-organic control involves, say, two or three sprays with a decent systemic spray such as Orthene or perhaps Confidor. If you use a water-based spray it certainly pays to add a wee bit of Conqueror Oil as well, to break down the waxy barrier of fluff so the active ingredient can have direct access.

I have found the best organic spray to be the oil by itself, but you do need a number of repeat applications. This will not only allow for a second and third hit, but it will also mop up those larvae that hatch from eggs at a later date. Good spray coverage on the plant is absolutely essential.

Palms and ferns and other wimpy plants do not react very well to mineral oil sprays — it really burns their leaves.

So what about the old trick of a fine paintbrush dipped in methylated spirits? All you need to do is touch the mealybugs

with the meths and they're history. No sprays needed — no stinky stuff — and there's the perverse satisfaction of seeing the hated mealybugs curl up and die.

But how good is your eyesight? It's all very well destroying the big ones we can see easily, but the first instar larvae are absolutely minute — you'll never see them without the aid of a hand lens. They are also numerous within a population and form the basis of the next, very successful generation on your plant!

Predators and parasites

Luckily there are also some great predators and parasitoids that will kill mealybugs — specialised ladybird species come to mind.

The steelblue version chomps on scale insects as well as mealybugs but the black and orange ladybird's job description is far more specific. It is called the mealybug ladybird (*Cryptolaemus montrouzieri*) and it has larvae that also prey on mealybugs. They devour the whole prey and recycle the white waxy materials to make themselves look just like their targets — a matter of wolves in sheep's clothing.

Parasitic wasps are also on the case when troublesome suckers vandalise your plants. Don't worry, the tiny wasps will find your infestations naturally, all by themselves. All you need to do is cease insecticidal treatments and plant some pollen and nectar-rich flowers for the adult wasps to feed on (see page 119).

Passionvine hopper

THE PASSIONVINE HOPPER (*SCOLYPOPA australis*) is an Australian species that probably made its way to New Zealand in numerous batches of eggs on plant material. This hopper is extremely common in the northern, warmer parts of the North Island, but has been known to be a problem as far south as Nelson and Marlborough.

It is notorious for its habit of jumping. When you brush past its host plant, you'll end up getting covered with clouds of nymphs, and in bad infestations breathing

SUCKING BUGS

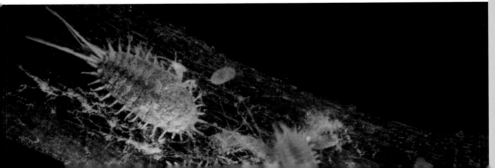

can become a problem — they're everywhere! Their fluffy tails, however, are quite handy for the flightless nymph as it has been suggested that these are used as parachutes to retard their rate of fall.

The passionvine hopper has well and truly made New Zealand its home. What's more, they manage to make an excellent living here on our native and introduced plants. In fact, they tend to be a heck of a lot more common here than in their original country of residence.

Life cycle

The species over-winters in the egg stage and goes through just one generation per year. The eggs hatch in spring as tiny little nymphs that I call 'fluffybums', a name that describes exactly what they look like: small, frog-like bug nymphs with an ornate tuft of fibres that look like a nylon tail

implanted in their bottoms (pictured below). From late January onwards the nymphs go through their final moult and turn into adult insects (passionvine hoppers), complete with wings. You can find these adults well into May. People also refer to them as 'lacey moths', but they have nothing to do with the Lepidoptera. They are true sucking bugs.

The adult hoppers are more mobile than the larval stages: not only can they still jump, but they now also have the power of sustained flight, albeit feebly. Reproduction is the key activity for any adult insect, and the passionvine hopper is no exception. After mating, the females dust off and sharpen their short but powerful ovipositors, and load them with cartridges (also known as eggs).

Eggs are thrust into the thin twigs and tendrils of the host plant. Every time the female withdraws the ovipositor from the twig, a little tuft of wood fibres comes out. These tufts are easily spotted, as they are situated at regular intervals along the

TOXIC HONEY

WHEN PASSIONVINE HOPPERS FEED ON THE POISONOUS SAP OF THE NATIVE TUTU (*CORIARIA ARBOREA*), THEY PASS ON THE TOXIC MATERIAL BY ADDING IT TO THE HONEYDEW. BEES THAT FORAGE ON THIS HONEYDEW INCORPORATE IT INTO HONEY STORED AT THE HIVE, AND ALTHOUGH TUTU SAP IS NOT TOXIC TO PASSIONVINE HOPPERS OR BEES, IT IS DANGEROUS TO HUMANS! THIS IS WHY BEEKEEPERS ARE NOT ALLOWED TO SITUATE HIVES IN THE VICINITY OF TUTU WEED PATCHES.

oviposition site, giving the twig or tendril a distinctly serrated look. The eggs start to become noticeable from March onwards.

Impact

As soon as the little nymphs hatch they will suck plant material through their tubular mouthparts. The sweet, nutritious phloem is where it's all at.

The excreta of the fluffybums is that sweet, sticky honeydew, an unmistakable diagnostic symptom of so many sap-sucking insects. They splatter honeydew everywhere with little regard for the aesthetics of the plant material below. A typical side-effect is the growth of a black sooty mould on the honeydew. Ants, wasps and bees are also regular visitors to the passionvine hopper's ablutions.

Substantial damage can be caused by both the fluffybums and the adult hoppers. Distorted growth of the plants they attack is merely one result of their mainlining activities; the deposit of honeydew and secondary sooty mould causes severe problems for growers of export produce. It is unclear as to whether the bugs transmit viral diseases from plant to plant, but the possibility cannot be ruled out.

Control

Judging from the number of queries I get about passionvine hoppers, it comes as no surprise that their control is just about impossible.

In the early weeks of development the nymphs are so small that no-one really notices them, yet this is the time when chemical or even organic control would work the best. A simple misty spray with pyrethrum or even synthetic pyrethroids would control the majority of the very small and still susceptible nymphs. When you try to spray them at a later, larger stage, you'll notice that they try to escape the toxic clouds by simply jumping away from the danger zone.

If you really want to use a mild spray, such as pyrethrum, it pays to be on the ball and hit the smallest, first instar nymphs in November. Just keep an eye on the new twigs of traditional host plants and blitz them as soon as they appear. A couple of repeated applications a week or so apart may be useful to cover the late-hatching populations of fluffybumlets. Heavier materials are the systemic sprays, which are taken up by the plant and contaminate the bug's food supply as well.

If you really abhor the use of an insecticidal compound, think about the weak link in the hopper's life cycle: the eggs. These are around for at least seven months of the year (March–October) and are easily detected once you know what to look for. Removal of these eggs will reduce populations of this pest for next spring. Every time you go out into the garden in autumn and winter, take your secateurs and cut off the series of tufted, egg-containing twigs and tendrils until you've got a good fist-full of these over-wintering ova. What you do then is up to you. It's best to chuck 'em into the fire-place. One thing is for sure, they're no good in the compost heap!

Parasites

The passionvine hopper's egg parasite *Centrodora scolypopae* will have to be active from March onwards if it is to be successful in maintaining its own populations. This minute wasp completes its life cycle inside the egg of the passionvine hopper, and is therefore a useful, tiny tyrant!

Psyllids

HERE IN AOTEAROA WE have some endemic pit-dwelling psyllid species that are well known to gardeners but are relatively harmless to their chosen natural hosts. The best example can be found making dimples on the underside of the leaves of pohutukawa and rata.

The insect *Trioza curta* will make itself at home only on these trees, and attacks no other plants. The damage only occurs on the smaller plants with juvenile leaves; psyllid nymphs will never settle on adult foliage, probably because the dense, fine, grey hairs on the underside of these leaves cause discomfort to the ticklish bellies of our little native sucker.

In early spring some years ago, I noticed some Acmena hedges in Auckland with an exorbitant amount of black sooty mould on the leaves. Sucking insects are often the cause of sooty mould and I found heaps of small nymphs attacking the new flush of growth on the hedge. Their shape and size reminded me of psyllids. A colleague identified them as a psyllid species, probably from Australia, or perhaps from the Pacific. However, it appeared to be an undescribed species.

The Acmena psyllid (larvae pictured below) belongs to a group with free-living larvae that produce copious amounts of white waxy threads from their anal glands, and lots of sticky honeydew, which makes a real mess. This sweet honeydew is a brilliant substrate for the black sooty mould fungus to grow on.

Life cycle

Psyllid larvae go through four or five instars, shedding their skins at regular intervals and becoming larger with each moult. The final instar changes into a fully winged adult psyllid, which resembles a sturdy aphid in shape and size. These adults also suck sap, but have the peculiar habit of leaping off the plant material when disturbed, before using their wings to fly

ack to the host. This gives them the cute
ernacular name of 'jumping plant lice'.

But when the adults engage in the
oble act of mating, they're not so jumpy,
h no! It is a pretty quiet and boring affair
nat results in the gravid female laying eggs
1 nooks and crannies on the new growth
f the plant. As long as the host plant is in
:tive growth, the life cycle of these
syllids will continue, and a number of
ibsequent generations can be completed
ithin a single growing period.

Psyllid species create intricate little 'bus
nelters', known as lerps, from the sugars
ney excrete. They use their excrement as a
uilding material in the most amazing way!

The psyllid *Eucalyptolyma maideni*
ictured below) builds pure white lerps.
hese little houses are about 6 millimetres
ing and look a bit like tunnel-shaped
loos. They stand out on gum foliage,
specially when the density reaches many
ozens per leaf.

Nymphs will quietly sit under their
rp, sucking away at the sap, for most
ionths of the year. Often you can see
iem poking their head out the opening as
to survey the environment. When
sturbed the greenish juveniles will
iickly run away over the leaf.

Other species create larval pits or
mples. These local leaf deformities are
the result of enzymes and toxins injected
into the leaves by the nymphs. All
juvenile stages will inhabit these dimples
before metamorphosis.

Impact

Like all suckers, psyllid larvae leave a large
mess. The host plant looks awful, covered
in honeydew and black sooty mould. No
doubt, some species will also act as virus
vectors.

Another free-living psyllid species
attacks the new leaves of boronia. This
insect, *Ctenarytaina thysanura*, is originally
from Australia and related to the Acmena
psyllid species. It's been here for quite
some time now, and has in the past few
years also been noticed damaging
Eriostemon.

The South American pepper tree
often shows the characteristic damage
patterns caused by yet another new psyllid
species, *Calophya schini*. This beast
probably arrived here via California. The
sucking activity of the juvenile, rather flat,
psyllids causes the leaves to make dainty
little dimples in which the larvae settle,
out of the wind and weather, and usually
on the underside of the leaves.

Cardiaspina fiscella is probably the
most damaging psyllid species on

eucalyptus in our country, as it is in Australia. It's been here since 1995, and the larval shelters are in the shape of a seashell. They have the consistency of lacy baskets with many little holes. Under these protective shelters the larvae go through their growth stages; the holes in the roof probably play a role in the regulation of relative humidity.

These psyllids prefer the new growth of the gum tree because that's where all the phloem sap is used. Older trees tend to suffer more from Cardiaspina than young trees. Attacked leaves show brown, dead patches around the lerps, and many leaves will be killed by the continuous attacks. This psyllid often causes severe defoliation and repeated events result in dieback of branches and weakening of trees.

In bad infestations of *Cardiaspina fiscella* dozens and dozens of lerps can occur on one leaf, usually on the underside. The other side of the leaf develops a reddish discoloration. Host species most affected appear to be *Eucalyptus botryoides* and *E. saligna*.

It is now distributed throughout Northland, Auckland, Coromandel, Bay of Plenty, Gisborne and Hawke's Bay. It appears to be active throughout the year, slowing down somewhat during the winter months.

Benefits

The toxins or enzymes injected by the young, sedentary nymphs of the native pohutukawa psyllid do not affect the health of the pohutukawa or rata. They've been living together now for many millions of years, and the evolutionary process has reached a nice balance between the sucker

BIOSECURITY

THE REGULARITY WITH WHICH NEW INVERTEBRATES SET UP SHOP IN NEW ZEALAND HAS BEEN THE TOPIC OF MANY DISCUSSIONS. QUARANTINE SYSTEMS ARE BREACHED WITH ALARMING FREQUENCY. THE NEWLY ESTABLISHED BEASTS CAN BE A NUISANCE TO HOMEOWNERS AND GARDENERS OR, IN THE WORST CASES, OF ECONOMIC OR ECOLOGICAL SIGNIFICANCE.

WHITE-SPOTTED TUSSOCK MOTH, PAINTED APPLE MOTH, SITONA WEEVIL AND CLOVER WEEVIL THE AUSTRALIAN SHEEP BLOWFLY, WESTERN FLOWER THRIPS AND VARROA MITE ARE JUST SOME EXAMPLES IN THE ECONOMIC-SIGNIFICANCE CATEGORY. EXAMPLES OF INVERTEBRATES WITH AN ECOLOGICAL IMPACT ARE ASIAN PAPER WASP, ARGENTINE ANT, AND THE SPRINGBOK MANTIS. GARDENERS BATTLE THINGS LIKE ORIENTAL FRUITMOTH IN STONE FRUIT, HIBISCUS BEETLE, SOYABEAN LOOPER (ON HERBS, ETC), AS WELL AS A FEW NEW DISEASES ON BUXUS, AND CAMELLIA PETAL BLIGHT.

ON TOP OF ALL THIS WE'VE HAD SOME BRUSHES WITH FRUIT FLY SPECIES, SNAKES, RAINBOW LORIKEETS AND SCORPIONS . . . WHEN IS IT ALL GOING TO END?

nd its host plant. When we take a
detailed look at this symbiosis, we must
ask the question: why do the pohutukawa
and rata help the psyllids by creating cosy
little dimples?

The answer can be found at the rear
end of the larvae. Native ant species patrol
the psyllids to collect any sugary waste
products that may become available and,
while doing so, they protect the psyllid
nymphs and the leaves they are living on!
Indeed, pohutukawa and rata leaves with
psyllids in residence hardly ever show signs
of chewing damage.

So, the psyllids get the phloem sap
for development, plus the assistance of the
host plant in the shape of those dimples.
In return for this sacrifice the tree gets
protection from the ants, which in turn
harvest the excrement of the insects. This
is a wonderful example of an ecological
ménage à trois, based on poo.

A similar system is set up between the
pittosporum psyllid (*Trioza vitreoradiata*)
and just about all species of this native host
genus. *P. crassifolium* often shows
particularly spectacular damage, especially
in spring, causing deformation of the new
growth and characteristic, almost arty,
crystals of white honeydew on black sooty
mould. Here again, in a well-established
shrub or tree, planted in the right place,
the damage is usually cosmetic and the
ecosystem requires no intervention by
the gardener.

Control

When it comes to organic or chemical
control of the introduced psyllid species,
the ones with the free-living larvae are
the easiest to control. Traditionally,
systemic insecticides have been used
effectively, but if you're more interested
in an organic approach, try some
pyrethrum or neem oil.

I do not encourage the control of the
psyllid on Acmena, as this insect does a
good job hassling that weed. I certainly
do not recommend any chemical action
for the control of our native psyllid
species. In fact, I would like to make a
plea for leaving them alone! And this also
goes for the thoughtless chemical control
of these symbiotic insects on native plants
for sale at garden centres!

Wouldn't it be a nice idea for garden
centres to sell young pohutukawa and
rata, as well as Pittosporum plants,
complete with their native symbionts?

SUCKING BUGS

Just stop spraying and write a simple, accompanying leaflet explaining that the gardener no longer merely buys a native plant, but a fascinating, totally endemic and organic ecosystem.

Predators and parasites

Predators (and, to a certain degree, parasites) take advantage of psyllid infestations, the most notable being the steelblue ladybird beetle and its larvae. To these predators the psyllid nymphs must be akin to having a smorgasbord of 'sitting ducks'. Predatory mites and silvereyes also take their annual toll.

In spring 1999 it became apparent that some kind of parasitoid was having a go at *Cardiaspina fiscella*. Neat little circular exit holes were found in the lerps of this psyllid, indicating that a tiny wasp had chewed its way out of its host. That wasp was identified in December 1999 as *Psyllaephagus gemitus* and it looks as if it is making a bit of an impact on psyllid populations. Hopefully this is yet another example of Nature carrying out its balancing act.

Scale

SCALE INSECTS COME IN a truly bewildering array of shapes, colours and sizes. Some are round, some build armour like spun-sugar stars, some are oval, while others are mussel-shaped. They can be black, brown, creamy-coloured or pure white; some have a waxy wig, fringed with red and black patches — a treat to watch under the microscope. And the nice thing is that scale insects usually sit very still.

Scales can be divided up into four groups: the armoured scales, the un-armoured scales (also known as soft scales), the felted scales, and the naked scales. All that counts is that some of them can cause serious damage to plants and crops.

Life cycle

The life cycles of the various species are rather confusing. Mature females either lay eggs or deposit live young under their protective caps or in their ornate egg sacs. The instar larvae or nymphs, known in scale lingo as 'crawlers', are the dispersal stage: they walk over plant material to a suitable spot to feed. These crawlers are so

minute and light that they can float in the air on wind currents, many miles away from the place where they were born.

After settling, the crawlers of some species (especially armoured scales) moult their skin and, with it, their legs, which makes them unable to move again. Other scale larvae retain the ability to move around until they are mature.

The sole objective of larval insects is to feed and grow, and scales do this by puncturing the host plant with their long, tubular mouthparts to extract nutrients from the phloem sap.

Most settled larval scales have a large number of glands or pores in their skin. These produce numerous fine, waxy threads that form the basis of their scale 'cap' or covering. In armoured scales, all these threads mat together and become a hard, waterproof shelter, detached from the insect's body. With each skin moult, the inhabitant enlarges its covering to accommodate its growing body and pushes the old, shrivelled-up skin into the cap. This then forms a different-coloured patch in or near the centre of the scale.

The unarmoured species (pictured opposite) have their cap as part of their bodies. The upper skin toughens and acts as the protective cover. Soft scales are round to oval and dome-shaped, with a shiny surface.

Naked scales are the true creative excreters of the insect world. They tend to protect themselves with a layer of waxy powder or filaments and most species remain mobile throughout their life.

Felted scales live inside their own cocoons of felt-like material. These capsules, also known as tests, are usually soft, but some species construct hard cases.

Impact

Greedy scale can be found on a range of woody hosts, but are especially a problem on export kiwifruit and feijoas. Other pestiferous armoured scales are the white rose scales (on the base of your rose plants; pictured below) and a similarly coloured species on the leaf blades of flax. The rose scale has male forms, and their juvenile stages can be readily identified as the smaller and narrower ones in amongst the round, large female scales.

Cymbidium orchids can sometimes harbour whole populations of fawn brown scale specimens in mixed sizes on leaves and other green parts.

SUCKING BUGS

Some of the larger, native, unarmoured species can cause distinct dimples in the leaf surface of their host plant; others simply smother the hosts and remove significant amounts of valuable carbohydrate from the plant's system, while covering it with honeydew and sooty mould. In severe cases plants show the results of scale activities by a serious reduction of vigour and health.

Control

Some scale insect species are a real problem and difficult to control unless you know the tricks of the trade. From a common-sense point of view it pays to cut out and burn affected branches as soon as you've discovered a scale problem.

If you must use a spray, first try simple, organic mineral oil. You can use it by itself, but remember to put on a number of repeat sprays — say four weeks apart — to suffocate the protected adults and kill the wandering crawlers and settled nymphs. The younger they are, the more susceptible they are, but eggs are generally immune. That's why repeat sprays are an important key to successful control.

Be aware, however, that certain 'wimpy plants' are not very tolerant of oil sprays. Soft-leafed plants and ferns, palms and orchids often show the phytotoxic effects of a mineral oil deposit. Check for sturdiness on one or two leaves before going in for wholesale slaughter!

Rose scales are easily dealt to in the middle of winter, just after pruning. Simply paint or spray the remaining stems with diluted oil or lime sulphur to soften the scales up a bit, then use a wire brush to remove them from the bark.

Some people mix the oil spray with systemic (organo-phosphate) insecticide, to increase the immediate kill rate by con-taminating the phloem sap as well. That's OK, as long as you realise that other, beneficial insects will also be hammered.

The general recommendation is to stick with a simple, oil-based concoction because all the scales' waxy coverings protect the insects from water-based insecticides.

Predators and parasites

The steelblue ladybird (*Halmus chalybeus*) is a great enemy of scales on flax (pictured

elow left) and citrus, so leave them alone
nd try to minimise insecticide use when
ou see them.

Native scales on native trees and
hrubs tend to be controlled by their
atural predators and parasites — and
here are heaps of different organisms
hat do that: kokako, silvereye, grey
varbler and other active insectivores
pend large amounts of time scouting
lants for small invertebrates, especially
cale insects. Predatory and parasitic
naggots of various fly species, as well as
arasitic wasps, also add to the biological
ontrol menu, as do certain ladybird
eetles and their larvae.

Greedy scale

GREEDY SCALE (*HEMIBERLESIA
RAPAX*) is a good example of an
rmoured species. If you carefully lift off
he hard cap, you'll see a plump, yellow

female stuck by its stiletto mouthparts to
the host plant. If you're lucky, you will
also find her eggs (sometimes a hundred
or so), or the hatched crawlers sheltering
under the maternal cap. There are no
male greedy scales as the females are able
to reproduce all by themselves.

Giant kanuka scale

MOST GARDENERS WILL BE familiar
with the masses of white cocoons that,
in some years, litter the bases of kanuka
trunks and other plants in autumn and
winter. These are the male 'pupae' of
the giant kanuka scale, *Coelostomidia
wairoensis*. The male (pictured below) and
female scales lead a remarkable life on the
twigs and branches of their host tree.

Females go through four stages: a
mobile crawler stage, two non-mobile

feeding stages, and finally a fully legged, mobile, non-feeding adult female stage. The males need five stages for their development from mobile crawler through just one non-mobile feeding stage followed by three non-feeding stages — one of which is those white cocoons — finally becoming a fully winged, tiny adult male.

Most people are alarmed at seeing their favourite teatree covered in masses of white cocoons and frequently demand a recipe for death and destruction. There is no need for that as this native scale is simply part of the ecosystem; besides, all you see in autumn is the end of the life cycle. Just watch for minute and fragile male scale insects flying about and crawling over the white, felty cocoons. It's truly spectacular. These little fellas are on the hunt for their female's tests so they can engage in a spot of R and R!

Sooty beech scale

THE SOOTY BEECH SCALE (*Ultracoelostoma assimile*) is particularly prevalent in the upper South Island beech forests, where they are the major source of honeydew for kaka and other nectivorous birds — that is, until exotic wasps started to compete for this valuable resource.

Sooty beech scales live a sheltered life in their tests on beech trunks, sucking sap through the bark of the tree. These scales have developed an interesting adaptation to prevent their habitat from clogging up with honeydew — a long, hollow, anal

filament made from wax directs the honeydew away from the tree trunk and presents it as a glistening droplet to anyone who might be interested in a sample. The result is an often dense mat of white and tacky hairs sticking out of beech trunks in the forest.

Giant shaggy scale

THE DARK, WET BEECH forests are also the home of the curiously named giant shaggy scale (*Coelostomidia pilosa*), a remarkably large and mobile insect in the adult female stage. Up to 10 millimetres long and resembling a soft, pink jube-on-legs, the female giant shaggy scale cruises over bark and mosses on the forest floor, waiting for a tiny winged male to alight on her abdomen so she can lay fertile eggs inside a granular capsule.

Cottony cushion scale

ICERYA PURCHASI IS THE exotic name given to the cottony cushion scale, a creature that can reproduce as a hermaphrodite (eggs and sperm produced by the same individual). The adults look smart in their grey, red and black upper surface, with a pure white, grooved 'skirt' of soft wax (pictured page 102, right). This is their fluted egg sac, containing hundreds of bright red eggs that hatch into tiny red crawlers with black legs. The crawlers escape the egg sac through the grooves and wander off to settle on the veins and midribs of the leaves of the host plant, to start the life cycle all over again.

Chinese wax scale

THE CHINESE WAX SCALE (*Ceroplastes chinensis*) looks comparatively drab in the adult form — greyish-pink, soft and amorphous blobs, lined up along the twigs of the host. However, when the crawlers emerge the picture is completely different. Hundreds of tiny, star-shaped, white nymphs (pictured below left) settle themselves along the veins of the host plant's leaves, where the phloem sap is available in good quantities. In heavy infestations the leaves can look as if they're covered with tiny specks of white dust or snow. As they develop, the nymphs become increasingly dark pink, but retain their white, waxy tufts for quite a while.

Spittle bug

ASK ANY ENTOMOLOGIST WHICH insects generate most questions from gardeners and growers alike, and chances are the list will contain spittle bugs (adult pictured below right) from South Island folk, and 'lacey moths' from North Island enquirers. Both species belonging to the plant sucking insect Order Homoptera, and can cause spectacular symptoms on a wide range of plants.

There are two species of spittle bug in New Zealand: the meadow spittle bug (*Philaenus spumarius*), and the variegated spittle bug, which goes by the wonderful and complex name of *Carystorterpa fingens*. Both species appear to like the cooler climates of New Zealand and have similar life styles.

Philaenus has only been in New Zealand since the early 1960s and it has caused quite a stir since then. This species has been reported on a huge range of host plants. Spittle bugs are very easy to detect, so the records are probably genuine.

Adult bugs are some 6 millimetres in size and have a characteristically blunt head and very obvious bulgy eyes. No

SUCKING BUGS

wonder this group of Homoptera is often called 'frog hoppers'. They are light brown in colour and look nothing like their larvae. One of the things that makes spittle bugs famous is their ability to jump.

Life cycle

Adult spittle bugs lay their eggs, in autumn, in neat little rows of a dozen or so. The bases of the eggs are enveloped in spittle, presumably to stop them from drying out.

As soon as the little green nymphs have hatched from their eggs in spring, they look for a good spot to insert their sharp hypodermic needle-like stylets. The aim is to hit a vein or vascular bundle in the plant, just as all suckers do. Basically, the spittle bug is trying to intercept nutritious sap from within the plant.

Older nymphs often show a very dark brown to black head, thorax, and developing wing-buds, which contrast sharply with the lime-green abdominal segments. The final moult changes the nymph, or larva, into a fully winged adult spittle bug.

It can leap and fly, and will do so when disturbed. While the adult meadow spittle bug tends to be a boring light brown colour, the variegated species at least has an interesting pattern on its wings.

Impact

Normal Homoptera vandalise the phloem, which transports the sweet carbohydrates made by photosynthesis. However, spittle bugs vandalise anything, especially the xylem, the principal water-conducting tissue of plants, as the excreted froth has no sweet taste at all. It's more like a green, mineral and slightly herbaceous flavour, and these tasting notes come from a wine aficionado!

It is this blob of spittle — sometimes referred to as 'cuckoo spit' — that alerts the gardener to the existence of the bug on plants. You simply can't miss it. When the bug feeds on the plant sap, it excretes the excess moisture and mixes it with air by vigorously moving its legs and abdominal body plates. This produces the frothy mess that acts as a protective cover for the developing sucker. Inside this watery shelter you soon find the nymph and, if you remove its spittle, you can observe it making more of the stuff (pictured below).

Despite the fact that the spittle bug belongs to that famous insect Order of 'flying hypodermic needles' there is no

idence that they are implicated in the ectoring of viruses and other diseases om plant to plant. In other words, it is oubtful if they can be labelled plant pests hen they occur in low densities. Even in igh numbers they may not cause as much ress as you would tend to think when you e the numerous and obvious patches of uckoo spit. Nonetheless, on some plants e sucking activity of spittle bugs can cause me distortion and dieback.

Control

you really feel that your plants are etting a hard time, it is still rather tricky control these perfectly hidden bugs. The andard recommendation is to hose off the ffending spittle with a good jet of water. The alternative is to lick the spittle off.) his exposes the nymphs for a wee while, nd allows a follow-up treatment with a ontact insecticide or even a systemic. But on't wait too long otherwise they'll have emselves covered before you arrive with e poison.

Hand-picking the bugs can be a ery satisfying job, especially if you have hickens, which are particularly fond of spittle bug canapé.

The citrus whitefly (*Orchamoplatus citri*; pictured previous page), first found in the late 1990s on Auckland's North Shore and Te Atatu Pensinsula, is a very recent addition to our fauna and it is spreading at a great rate. Citrus whitefly tackles the whole range of citrus in my backyard — lemons, kaffir limes, mandarins, oranges and limes — and colonises the undersides of the fresh, young leaves with gusto.

Cabbage whitefly (*Aleyrodes proletella*) is an old faithful. It's been here for yonks and most growers of cruciferous vegetables should know this pest well, as it covers brassica leaves with sooty mould and white specks. It is most numerous in late summer, but hangs around well into autumn in the northern parts of the country.

Life cycle

Greenhouse whiteflies can go through their life cycle quite quickly and this, coupled with a healthy resistence to insecticidal sprays, can cause the familiar outbreak situations.

Eggs are laid singly on the newest leaves. They are creamy in colour and hatch in about a week or two. The small nymphs look a little like flat, transparent scale insects and these suckers settle themselves on the leaf with their mouthparts firmly embedded in the rich, sweet veins of the vascular system.

As the nymphs grow older and shed their skin, they start to develop a characteristic glassy fringe of scales on the perimeter of the body. Larval development (usually four instars) can take about a month to complete.

The adults hatch from the last instar (and non-feeding) nymph stage. These white-winged small insects will congregate around the tops of the host plants, where the newest and freshest leaves are the prime oviposition sites for the start of the next generation. Greenhouse whitefly are able to complete many generations per year, as long as there is a continuous availability of host plants and plenty of warmth.

The ash whitefly and cabbage whitefly are not quite as prolific with their number of generations per year and the new citrus whitefly appears to go around the generational track just once a year. But that does not make them less of a pest!

Impact

The signs and symptoms are quite clear as

hese insects remove substantial amounts
f carbohydrate from the host plants,
aving a distinct debilitating effect often
ccompanied by extensive yellowing
f the leaf tissues. There will also be
oneydew deposits, which will be
overed, in time, by sooty mould.

Cabbage whitefly is easy to spot on
he underside of its vegetable host — the
emi-circles of eggs stand out like a
owboy camp on the prairie.

The citrus whitefly causes a heck of a
ot of sooty mould on the upper sides of
he leaves (pictured opposite left). A lemon
r mandarin tree will soon turn black.

Control

Greenhouse whitefly can hammer a plant
o death, if left uncontrolled. Unfortu-
ately they tend to be rather tolerant of a
vide range of insecticides — even the
eavy-duty systemics are often no longer
fficacious, nor is the insect growth-
egulator Applaud the panacea it once was.

Organically speaking, some folks are
inding that fatty acids (potassium salts) and
eem oils will do the job as long as they're
pplied on a regular basis. Have a go! I
uggest you tend to direct your sprays to
he underside of the leaves.

Yellow sticky traps, hung above the
vhitefly-infested crops, can catch many a
lying Trialeurodes. It might sound a crude
method of control, but when well-placed
near their natural oviposition sites) these
raps will do damage to adult populations,
especially in windless conditions such as
nside glasshouses.

Predators and parasites

On crops grown outside, some predators
may be of benefit. When examining a leaf
with a good whitefly infestation with a
hand lens you'll often find small, bright
orange maggots creeping about the
pestiferous wildlife. These are more than
likely the predaceous larvae of a small
Cecidomyiid fly (see page 85) or, more
correctly, a midge. Anything within their
size-range is fair game, and the predatory
devastation can be quite considerable.

For those of you who can't help but
grow whitefly in the greenhouse, one of
the world's most famous biological
enemies can be bought by mail order and
employed in the battle against your
whitefly pest: *Encarsia formosa* (pictured
opposite page, right). This tiny wasp is the
most effective parasitoid in the control of
whitefly on greenhouse crops. Infected
whitefly larvae turn black as Encarsia
develops inside them. This species is
commercially available under various trade
names in New Zealand, and is probably
the most frequently sold insect in the
world. It needs to be re-introduced from
time to time as the parasitoid's population
tends to fluctuate throughout the seasons
— most commercial growers just release
thousands of these wasps into their glasshouse every month or so,
using these parasitoids almost as a
regular application of biological
control. Unfortunately, Encarsia is not
very efficacious in the big, bad world
outside.

Ants and wasps

Ants

ANTS ARE LITERALLY EVERYWHERE. We think there may be as many as 20,000 species. In Australia there are possibly some 3000 species and most have not been described. In New Zealand we are looking at a mere 40 species, with at least three-quarters of those being introduced.

Impact

It's the introduced species that nest inside your wall cavity or roof and send huge marauding trails of workers down to your kitchen bench, pantry, pet-food bowl, or even bathroom. And they drop large amounts of insect body bits and other debris from tiny holes in the ceiling; all over your clean, white bed linen.

There are some ants that rate highly on the human scale of nuisances. The white-footed ant (*Technomyrmex albipes*; pictured opposite, left) is probably the best known in the urban areas of New Zealand. Originally from Asia, it loves to live in the warmer climatological regions, but can sustain itself nicely in heated buildings in Dunedin — and why not? If you are a recycler par excellence it pays to associate yourself with the messiest mammal in the world.

The white-foot is famous for suddenly appearing from nowhere in large numbers. It's also famous for its defence smell — a mixture of formic acid and other abdominal secretions that will taint even the newest hot-water jug when a few ants get boiled by accident. Many people will also become aware of this critter when light switches start to crackle and fail, often accompanied by the smell of burning insect flesh! If just

one ant happens to get trapped between the two electrical plates inside such a switch, the 240 volts plus a few amperes will frizzle the poor soul to smithereens. As it dies, the ant will release a huge cloud of 'help-me' gas, which stirs all its mates to come to the rescue . . . and, of course, the next ant will be electrocuted in exactly the same way, as it tries to pull its still-sparking companion from the jaws of the predatory switch plate. After a few weeks all these dead live-indicators will really start to stink!

Benefits

If we concentrate on our nine or ten native species we find that they seldom, if ever, enter our houses and do not have a huge nuisance value. They are, in fact, primitive species that act as generally useful recyclers of decaying organic materials, such as proteins and carbohydrates.

Ant activity on plants is, therefore, an easy-to-spot, live indicator of the presence of sap-sucking insects — the ones that can really damage your crops or ornamentals. As such I see these social Hymenoptera as a useful force in the outdoor environment.

But ants do not solely go for carbo-hydrates and sweet honeydews. Some species 'specialise' in savoury materials and recycle proteins. In the complex world of ecology these activities are part and parcel of the no-waste society of insects.

It must be stressed, however, that to simply divide the ant species up into 'sweet feeders' and 'protein feeders' is a wee bit simplistic, as most species can switch their diets and preferences according to temperatures, the season and the developmental stage of their colony.

Control

There are only a few occasions when it is necessary to control ants in the garden and outdoor environment. Nesting activity under paving stones often results in the excavation of copious amounts of paving sand, sending the bricks, tiles or stones out of level and alignment. Pouring liquid insecticides or kerosene into the entrance holes of the nest will move them on, but this may cause all sorts of unwanted side effects on non-target species too. Crushed naphthalene or camphor sprinkled around the ant's trails (pictured below right) and nests, will upset them to such an extent that they will move their home and offspring to a more suitable location.

ANTS AND WASPS

One of the most common questions I am asked relates to the control of ants inside the house. The simplest way is to get some residual insecticide — permethrin is a relatively safe active ingredient — in the form of an aerosol. All you do, however, is temporarily cancel the ants' access to the sprayed surfaces. As soon as the insecticide has been broken down by ultraviolet light, the ants can happily wander back.

Ant baits are probably a more lasting solution. The bait, which can be sweet, protein-based, or both, is laced with a very slow-acting insecticide, such as borax and boric acid. The idea is that the ants will carry the bait back to the nest and feed it to their larvae and queen, causing wholesale slaughter in the colony after a few months.

It can indeed take a long time to get on top of a nest with baits, but as long as you know how to play the game you should see some success. First of all, place the bait inside a pottle with a 5-millimetre hole in the lid. This prevents the liquid bait from drying out too much, keeping it attractive to ants for weeks and weeks.

Keep the bait stations out of the sun and place a number of them around the house, bearing in mind that you could have various nests in your house and that each nest has a very distinct territory. One bait station for each nest would be a good start.

Finally, you'll need quite a bit of patience and a commitment to keep the house clean, so that any alternative food for the ants is removed. And never use insecticides in association with a baiting strategy, otherwise the ants won't be able to get to the baits at all. Sounds reasonable, eh?

But please, always be careful what you are targeting with your baits or insecticides. After all, most ants are rather useful as recyclers and live indicators of troublesome plant pests! And they perform this service free of charge, round the clock, seven days per week!

HARVESTING HONEYDEW

IF YOU OBSERVE AN APHID COLONY FOR HALF AN HOUR OR SO, YOU'LL SEE THE INTERACTION BETWEEN APHIDS AND ANTS. THE LATTER WILL MANIPULATE THE APHIDS' BEHINDS TO ENTICE THEM TO EXCRETE MORE HONEYDEW — 'FEED ME, SEYMOUR!' ANTS THAT VISIT PSYLLIDS, OR 'JUMPING PLANT LICE', ALSO APPEAR TO BE HARVESTING THE HONEYDEW WITH GREAT GUSTO, ESPECIALLY AS THE SUGARY EXCRETA IS OFTEN CRYSTALLINE IN FORM AND HENCE EASIER TO CARRY BACK TO THE ANT'S NEST.

Argentine Ant

IN JANUARY 1990, ANT expert Olwyn Green was playing her viola at Auckland's

Mount Smart stadium during a rehearsal for the opening ceremony of the Commonwealth Games when she noticed a few unusual ants on her sheet music. The ants were subsequently identified as Argentine ants (*Linepithema humile*). This was a newly introduced species to New Zealand and it was found to already be well established in the area.

It is now the most serious ecological threat in New Zealand. It displaces all other ant species from its territory and preys on a large range of invertebrates, such as millipedes, spiders, beetles etc. It is also shown to have an impact on bird populations, simply because of the numbers of ants that are involved in attacks: birds are driven from their favourite feeding areas or breeding sites and young nestling birds are often killed and transported in tiny bits back to the ants' nest. In dry regions overseas, the Argentine ant is even known to cause chewing damage to the fibrous root systems of some plants, and it loves to live in regularly watered gardens.

Around 2000, this South American threat became established on Tiritiri Matangi Island. This was enough of a concern for the Department of Conservation to mount an eradication campaign that appears to have wiped out the species on the ecologically sensitive island.

Paper wasp

THE ASIAN OR CHINESE paper wasp, *Polistes chinensis* (pictured below), is a thin-bodied wasp. It is predominantly black with thin yellow lines on the abdomen, some yellow dots here and there, and reddish-brown legs, which are normally held dangling under the body when in flight. The wings are a smoky colour.

It is a close relative of the Australian paper wasp (*Polistes humilis*), sometimes referred to as the Tasmanian paper wasp. This beast is a rusty red-brown colour with patches of black on its thorax and abdomen. It too flies with dangling legs and smoky wings.

Paper wasps are not native to New Zealand. However, there's some evidence that the Australian variety has been here for more than 100 years and probably arrived in Northland as a stowaway on cargo carried across the Tasman by ships.

The Asian job is a more recent interloper. First sighted on the Whangaparaoa Peninsula in the late 1970s, it quickly spread to Auckland and beyond. They now occur as far south as Nelson and Marlborough, but are more numerous in the warmer regions of the North Island.

Life cycle

In early spring the pregnant female paper wasps start their nest-building activities. They chew wood fibres off fences and dead twigs and masticate them to construct the first papier-mâché cell of a new nest. In our garden, the nest sites are usually on the vertical face of the fence and pointing in a northward direction.

The busy queen then constructs a few more cells next to the first one, resulting in a golf ball-sized dwelling, attached to the fence by a short stalk. As soon as a cell is completed the queen deposits one egg deep inside, while she gets on with building the next cell.

In mid-spring (October), I once counted at least 10 proto-nests on our fence, of which only six were occupied. It appears that at that time of the year quite a few new nests are abandoned, possibly because of fluctuating weather patterns, but also because of queen mortality.

The eggs in the viable proto-nests hatch in their private brood chambers, and the resulting larvae are fed by their mother on bits of caterpillar and other invertebrates. Judging from the feeding behaviour of the establishing queens, the diet may include pollen, nectar and honeydew, plu the odd bit of vegetable matter or fruit pulp. When the larvae are fully grown, they pupate in their chambers, which are then closed off with a whitish cap.

The first brood of a dozen or so daughters is essential for the queen's surviv and the development of the nest, as these offspring will be the start of a new work force. They are going to continue extending the nest, while the queen will concentrate more and more on egg-laying. Slowly, during summer and early autumn, the nest will grow from a few dozen cells t an umbrella shaped condominium, with a couple of hundred cells.

Towards the end of summer the nest produces male wasps, which are characterised by their distinct yellow faces. These males are not concerned with the day-to-day running of the household; instead, they select a certain hedge or shrub as their display site to attract young queens for a bit of R and R. With winter looming, there's need for fertilised queens, and the life cycle gets into its hibernatory phase. The nests ar deserted, the male wasps die, and overwintering queens can be found under rock pieces of timber, or in the folds of your curtains.

Impact

The impact of paper wasps can be significant — try brushing against a nest and you'll know all about it! The residen wasps will raise their wings in a threat or alert posture, before flying off to deliver a rather painful sting. The problem is that i shrubs or bushy environments the nests are wonderfully camouflaged — you simply won't see them until you've been attacked by the irate occupants.

There is very little numerical evidence of the ecological impact of these exotic predators on our native invertebrates, but it would be reasonable to assume that they'll seek and destroy good numbers of native species in the process of raising their brood. In any case, as introduced organisms they make a thorough nuisance of themselves.

Benefits

f you observe these predators hunting for caterpillars on your cabbages, tomatoes and beans, you can not help but be impressed by their ability to find even the smallest pests, and by their ruthless method of cutting up large prey into transportable-sized bits of protein. They keep coming back until the whole damn caterpillar has been carried to the nest — and then they'll keep coming back to find some more victims.

The wasps appear to predominantly hunt by vision. There's no doubt that their range of prey species includes most invertebrate pests on our prized plants: soft-bodied larvae, vegetable bugs, flies, passionvine hoppers, insect eggs, and even species of soft scale insects. But those of you with the odd swan plant in the backyard will be well aware that paper wasps do not care if their food items are pest invertebrates or beneficial species — they'll take whatever they can lay their six little tarsi on, and that causes problems.

Control

Kill over-wintering queens immediately when found.

Established nests are easy to terminate at any stage. All the wasps are home after sunset and later in the evening temperatures are lower, so the reaction time is longer as they sit — head first — in their cells. Approach is easy and should be swift — either spray the face of the nest thoroughly with an aerosol (fly spray will do) or sever the nest's stalk, dropping it into a plastic bag (without holes). Tie the bag up before the dozy buggers back out of their cells, and chuck it into the freezer overnight for a cool kill.

Sawfly

THE SIREX WOOD WASP belongs to the Order Hymenoptera, together with the ants, bees and wasps. More accurately Sirex could be described as a sawfly, or plant-feeding wasp.

Impact

This interloper from Europe has been established here for more than 100 years now and its main targets are Pinus species, although spruce and larch are also on the menu, especially when these show signs of stress or decline.

The cunning technique used by the female Sirex wood wasp is to inject a bit of mucus and some fungus into each hole that she drills with her ovipositor. The mucus and fungus help to weaken the wood so that the larvae have no problem getting established. Eventually, the sapwood dies as a result of the multi-discipline attack. (Another species, the willow gall sawfly, is pictured below left.)

Control

In commercial forestry plantations control of this pest can be achieved by a mixture of good silvicultural practices (i.e. by reducing moisture stress and by maintaining the trees' vigour) plus a reliance on a number of rather effective biological control agents in the form of parasitoid wasps and even nematodes.

Parasites

A spectacular, large Ichneumon parasite (*Certonotus fractinervis*) has a very cunning way of locating and accessing candidate Sirex grubs in their wooden tunnel. The female's 60-millimetre long ovipositor is equipped with 'saw-blades' that allow the parasite to literally drill a hole from outside the branch to the tunnel inside. The saw-blade-tipped sheaths form a tube through which the ovipositor can be guided towards the hapless Sirex grub.

German wasp and common wasp

SWIMMING POOLS ARE WONDERFUL things. They not only collect myriad drowned invertebrates, but also attract the attention of German and common wasps. In summer, the big black-and-yellow Hymenoptera with the thin waist and short temper settles itself on the edge of the pool to drink.

The two Vespula species in New Zealand — *V. germanica* and *vulgaris* — are very similar to each other. Only the females (queens and sterile workers) can be reliably identified by their markings. The males often need to be identified by looking at their genitalia under the microscope.

The German wasp (*Vespula germanica*; pictured below) arrived in New Zealand during the Second World War in a consignment of aircraft parts destined for Hamilton. The common wasp (*Vespula vulgaris*; nest pictured opposite) is a much later arrival and, to be honest, we cannot be sure when it reached our fair land. Because of its similarity to the German species it was not 'discovered' until early in the 1980s.

Both species are now well established and occur throughout the country, from suburban gardens in the main centres to the beech forests of the South Island.

Life cycle

Hibernating fertilised queens wake up in the warmer spring weather and start constructing small, golf ball-sized nests, the 'queen's nests', to raise the first generation of sterile female workers. Nest sites vary — often it is a small hole in a clay bank or in soil, but hollow tree trunks and wall cavities are also very suitable, especially if handy do-it-yourselfers have put ventilation holes in their brick constructions.

Nest material looks like grey (germanica) or brown (vulgaris) papier mâché, and the bad news is that, more often than not, it has been recycled from the wood fibres of your deck or weather-boards. The workers chew these fibres, and repeatedly revisit rough patches of timber. The masticated wet wood is then moulded into quite a pretty, textured wallpaper.

As more workers are produced the nest grows quickly and adjoining soil is excavated to make way for extensions. As it's summer the workers visit swimming pools to take on board a cargo of water, which is transported to the clay nest area underground. The clay is moistened with water to facilitate further excavation.

The wasps literally scrape the clay with their mandibles and roll it into small pellets that can be clutched by the legs and flown out of the entrance hole. Once outside the flying wasp drops the pellet of clay and heads for the swimming pool again. Concerned householders often ring entomologists with queries regarding the sudden appearance of hundreds of small clay pellets scattered over car bonnets and bay windows. Here's your answer — you're under the flight path of a colony of Vespulae involved in excavation work.

In late summer and autumn fertile male drones emerge from nest cells, as well as large, fertile females that are destined to become queens. The timing is perfect, because the drones ensure that the queens will be fertilised before they hibernate.

In winter, most nests perish, but the odd one keeps 'ticking over', especially when winter temperatures are not severe. These nests have a head start the next spring, with an already numerous work-force. Often the over-wintered nests will grow out of all proportion during the second summer, sometimes reaching record-breaking dimensions with millions of cells. A particularly large nest found in

ANTS AND WASPS

the Waitakere Ranges measured 4.5 metres by 1.5 metres by 0.6 metres — not something you'd want to bump into on a pleasant Sunday afternoon tramp!

Wasp densities fluctuate from year to year and from district to district. Landcare entomologists have been studying wasps for more than a decade and found nest densities in the West Coast and upper South Island varying from five to 16 per hectare, with a maximum of 50 nests per hectare in bad areas in a bad wasp year! This last figure equates to five very grumpy colonies on a single quarter-acre section. With wasps in such numbers it is no wonder they impact on economy and ecology.

Impact

In spring and summer Vespid wasps hunt insects of all sorts and descriptions. Caterpillars, grubs, flies, bugs and even large dragonflies are captured, cut into sections and transported back to nests to feed the hungry wasp larvae. The sheer numbers of wasps on the wing in summer will give you some sort of idea of their predation potential, and most of their prey, especially in native bush areas, is undoubtedly of native origin.

In late summer and autumn the Vespula species switch diet from proteins to carbohydrates — from meat to chocolate bars. In our urban environment they'll go for half-empty bottles of soft drink and sweet cakes, but in the bush they'll be on the lookout for honeydew.

The horror stories from the South Island beech forests tell it all — specific scale insect species on the beech trees excrete copious amounts of honeydew during the course of summer and autumn, and our nectar-feeding birds, such as kaka, bellbird and tui, rely on this high-energy boost to gain 'condition' for the harsh winter. The healthier these birds are, the better they survive the winter, which results in a higher breeding success the following spring. Wasps have now been demonstrated to seriously compete with kaka for honeydew, causing a decrease in breeding success and hence a general decline in the kaka population.

Wasps also affect our economy. Tourists are generally not amused by hordes of stinging Hymenoptera, eager to pinch the ham or jam from their sandwiches. Wasps killing livestock is another story that hits the headlines from time to time, and fruit growers will be able to tell you all about the damage done to grapes, peaches and other fruits. And then there is a claim, involving hundreds of thousands of dollars worth of annual damage to the beekeeping industry, that wasps invade hives, rob honey, and often exterminate or seriously affect numerous honey bee communities. It's a mess out there when wasps do their thing.

Control

Kill the big over-wintering queens.

Locate the entrance of an established nest, and come back late in the evening with an accomplice and poison. You can use kerosene or diesel — you'll need at least half a litre, in a bottle. You can put

he opened bottle upside down into the
hole if the hole goes straight into the
ground — the fumes will kill the wasps
and the bottle blocks up the hole.

Powders can be used and registered
dusts such as Permex Insect Dust and Yates
Wasp Killer Dust are good alternatives.

For horizontal entrance holes it may
pay to use one of the dusts according to
the instructions on the pack — the idea is
to chuck the dust into the entrance so the
wasps fly through it. They can't help
getting contaminated and because of their
frenzied movements invariably distribute
traces of the dust right through the nest.

Warning:

DON'T light the kerosene fumes. The
explosion won't kill the wasps (the fumes
will), and your insurance company may
not be very impressed with you either!
DON'T use a garden hose with water to
flood' the wasps out of their cavity.
DON'T linger near the nest site after a
poisoning exercise in the evening —
frenzied wasps will be attracted to the
torch-light!

Parasites

Biological control involving a tiny parasitic
wasp, *Sphecophaga vesparum vesparum*, has
been tried since 1987, but with limited
success. The parasite was quite capable of
establishing itself in the wasp nests in
which it was released, but needed a taxi to
go from one nest to another. Its dispersal
was slow, to say the least.

More recently, a second subspecies of
this parasitoid, *Sphecophaga vesparum burra*,
was released in South Island beech forests
in the hope of a more aggressive attitude,
and a third subspecies, *S. v. israelensis*, is
also under field observation. With a bit
of luck they will get the idea and rip
through the Vespula nests with abandon.

Parasitic wasps

IT IS OFTEN DIFFICULT to explain to
growers and gardeners that there are
beneficial wasps in this world. The Order
of wasps — Hymenoptera — is synony-
mous with yellow and black aggression
and venom. However, within the Order
there are about a dozen or so families of
useful parasitoids to look out for.

Ichneumon wasp

THESE SIZEABLE INSECTS HAVE
striking colour patterns — orange-red or
yellow, with black patches and sometimes
white spots. In typical parasitic wasp-like
fashion they are often found cruising
plants, busily drumming their antennae to
pick up the scent of a possible host.
When a suitable specimen is found, the
female will insert her ovipositor into the
host and lay an egg. Some parasitoids lay
their eggs on the skin of the host.

Ctenochares bicolorus is a nice example
of a colourful Ichneumon wasp (pictured

below left). About 25 millimetres long, this orange wasp has a black abdominal tip, as well as black patches on the side of the thorax and head, and black-tipped yellowish wings. The antennae have a light yellow section about halfway down. Originally from South Africa, it became established in New Zealand quite some time ago and is often found parasitising the chrysalis of the green looper caterpillar.

The white-spotted *Echthromorpha intricatoria* is a smaller, native species of pupal parasite. It ranges in size from about 10–15 millimetres and is black with a number of white spots down its abdomen and thorax, with orange antennae. This Ichneumonid has a wide host range, including green loopers and tomato fruitworm, magpie moth, case moth and both admiral butterfly species.

Glabridorsum stokesii hunts out the pupae of leafroller moths and lays an egg on the outside of its host. The wasp's grub then tunnels its way into the chrysalis.

Braconid wasp

GENERALLY SMALL, BLACK WASPS, the largest species may barely reach the size of an ant. There are some formidable caterpillar parasites within the Family Braconidae. *Cotesia glomerata*, which was introduced from Europe and America to control the cabbage white butterfly, is a spectacular example. The cabbage white caterpillars are parasitised when they are very small and, despite their size, the female wasp manages to insert a couple of dozen eggs in each white butterfly larva

The eggs hatch into gregarious, but hungry, little wasp grubs which somehow keep their host alive; the irony is that a parasitised white butterfly caterpillar not only lives longer than a non-parasitised one, but it also eats more cabbage foliage.

When the host is fully grown, all those little devils chew through the caterpillar's skin and vacate the dying corpse. They then construct whole rafts of characteristic, yellow silken cocoons in which pupation takes place.

Pteromalid wasp

WHILE WE ARE ON the topic of hassling cabbage white butterfly, *Pteromalus puparium* contributes significantly to the reduction of this pest. This black wasp is a

few millimetres long and specialises in ovipositing in the chrysalis of the white butterfly, but also in those of the monarch and red admiral. A couple of dozen wasps develop within one chrysalis, and when these hatch from their host, they chew a neat little round hole in the chrysalis' skin (pictured opposite right). Pteromalus has synchronised its life cycle with that of the white butterfly, so it is an efficient parasitoid.

Aphid parasites

IF YOU HAVE A good look at your aphid populations on the new growth of plants, you'll often find some individuals that stand out. They are bloated, with a parchment-like, light brown skin. It is hard to imagine that a wasp can complete its life cycle inside the small body of an aphid, but that is exactly what happens. When the adult wasp hatches from its mummified host, it leaves a circular exit hole in the skin. To be honest, it is easier to detect the end result of parasitism in the garden than the parasitoids themselves.

Aphidius species are tiny wasps (pictured below) that insert an egg in a living aphid. *A. colemani* is such a parasitoid, and this species is now bred in captivity for sale to growers as a biological control agent against a range of aphids that damage crops.

Aphelinid wasp

THERE ARE A LOT of useful species within this Family of minute insects. *Aphelinus mali* is the parasitoid that single-handedly keeps the woolly apple aphid under control. *Centrodora scolypopae* is the wasp that parasitises the eggs of passionvine hoppers, providing an excellent example of being able to live in cramped conditions. But the most famous Aphelinid in the world is probably *Encarsia formosa* (see page 109).

Egg parasitoids

THERE ARE SOME WONDERFUL specialists in these groups. *Trissolcus basalis* (Scelionidae) is an introduced wasp that completes its life cycle inside the eggs of green vegetable bugs and other Pentatomid bugs. Green veggie bugs lay their eggs in large clusters, and when a

Trissolcus female finds an egg raft most, if not all of the eggs will be parasitised.

A few species of Trichogramma wasps occur in our country and their speciality happens to be leafroller eggs, including those of codling moth and Oriental fruit moth. Usually whole egg batches of leafrollers will be parasitised and these tiny egg injectors can be quite important as biological control species.

One of the craziest stories involves *Litomastix maculata* (Encyrtidae), an introduced polyembryonic parasitoid of the green looper. The adult female wasp lays just one egg inside a green looper egg. The wasp egg then divides and produces 1000–1500 larvae which kill the fully grown green looper larva, just before the latter is about to pupate. Parasitised green looper caterpillars are easy to spot — they are usually situated inside a half-finished silken cocoon and look grossly distorted, with thousands of little wasp grubs or pupae visible through their translucent skin (pictured below). Keep those green looper mummies in the garden, as the tiny adult wasps will hatch about 14 days after they have killed their host. Towards autumn a large percentage of the green looper population may be parasitised.

Moths and butterflies

THE WORD LEPIDOPTERA IS made up of lepidos, meaning 'scales', and ptera, which means 'wing'. The wings of our moths and butterflies are covered by thousands and thousands of tiny, flat, over-lapping scales (pictured below left) — a little bit like the tiles on a roof. These scales refract light in a bewildering array of colours.

The heads and bodies of moths and butterflies are densely clothed in hairs. There's no doubt that this fur keeps the beasts warm at night and minimises heat loss. This is very important when you rely on external temperatures for your living, moving and locomotion. The scales on the wings may serve a similar purpose, although these may also aid in smooth aerodynamics during flights.

And then there are the palps (projecting sensory organs — a little bit like external taste buds), a long, coiled proboscis or tongue (pictured below right), antennae (as a general rule, feathery and bushy in moths, clubbed in butterflies), eyes with many facets or ommatidia, six legs on the muscular thorax, and an abdomen full of digestive and reproductive 'hardware', which isn't really hard at all, but soft and squishy.

Moths and butterflies were invented about 130 million years ago, approx-imately at the same time as flowering plants, so it is not surprising that these two groups of living organisms are still inextricably linked. Like it or not, they depend on each other — the lepidopteran larvae feed on plants, and the nocturnal moth and diurnal butterfly visit flowers for a nectar feed, pollinating the plants as they go.

There are — at last count — about 150,000 different species of Lepidoptera in the world. One per cent of them — about 1500 species — are found here.

MOTHS AND BUTTERFLIES

Bag moth

PEOPLE ARE ALWAYS INTRIGUED to find the cigar-shaped bag-moth cocoons (pictured below) hanging from a shrub or tree, especially when there are dozens of them on one host. What's more, they never seem to move, yet somehow the plant shows signs of chewing damage in the form of holes and missing foliage.

To be quite frank, a lot of people never see the bag moth's larval cases at all, because of their cryptic grey coloration and the camouflage provided by bits of dried host plant material incorporated in the surface of the cases. They are masters of blending into the background.

You can find the cases on a wide range of hosts — they love ornamental conifers, such as Cupressus and Larix, but will devour leaves of willow, flax, titoki, kanuka, manuka, elm, acacia and Washingtonia palms with just as much gusto. You never know what plant in your garden will be next on the menu!

Life cycle

When the eggs of the bag moth (*Liothula omnivora*) hatch into caterpillars they use their silk to weave themselves a case that fits around them like the proverbial glove. Actually, it's probably more like a sleeping bag in shape and construction, wider at the top and narrow at the pointy bottom end.

The caterpillars are well equipped to hang on to their cases — the stumpy abdominal legs (known in the entomological trade as 'false legs' or 'prolegs') have

PURIRI MOTHS

LOOK FOR THE WOOD-BORING AND BARK DAMAGE-PATTERNS MADE BY OUR MAJESTIC, NATIVE PURIRI MOTH. LARGE TUNNELS INTO THE WOOD ARE USUALLY CONCEALED BY A DIAMOND-SHAPED PATCH OF SILK, COVERED WITH ALGAE AND LICHENS, SO THEY BLEND IN WITH THE TREE'S BARK. IN THIS TUNNEL, OUR MASTER CHEWER LIVES FOR ABOUT SEVEN YEARS, SCRATCHING AWAY AT THE CAMBIUM AS THIS LAYER RE-GROWS. IF THE HOST TREE IS SMALL, THE CATERPILLAR'S ACTIVITIES CAN RINGBARK IT, BUT BY THEN IT'S TOO LATE TO ACT.

OLD LARVAL TUNNELS ARE EAGERLY SNAPPED UP BY TREE WETAS AS SECOND-HAND DWELLINGS. AN EXAMPLE OF THE BRILLIANT SYSTEM OF RECYCLING IN THE INSECT WORLD!

umerous sharp, hooked crampons that fford a good grip on the silken interior ning of the bag. This leaves the three airs of thoracic, true legs, which are ituated just behind the head segment, free o grab hold of twigs and foliage.

Generally the caterpillar is very well rotected inside its case and it has some emarkable tricks up its sleeve. For example, t the top end of the case the inhabitant nakes a drawstring contraption, so that when a predator or human hand disturbs the eace, the caterpillar simply pulls the string ight and closes the entry. Clever stuff!

For many months the bag-moth larva ;oes about its business, just like any other aterpillar, feeding and resting, moulting nd growing, until it's time to pupate. By hen the cases can be up to 80 millimetres ong. Often an extra lining of fresh silk is dded to the inside of the bag, so that the upa or chrysalis will be nicely padded.

The transition from caterpillar, via upa, to adult moth takes place inside the ase, which makes sense, as the insect has nvested a lot of energy and material in onstructing this burglar-proof home. Might as well use it to the max, eh?

When a male pupa hatches, the moth rawls out of the bottom end of the case. t is a rather nondescript kind of moth, urry and black, with four relatively short vings, and a body length of perhaps 20 nillimetres or so. They fly very fast and rratically — a bit like an oversized fly — especially when they pick up the heromone scent of a female moth nearby.

When the adult female emerges from er pupal skin, she is hardly recognisable as

KAWAKAWA CATERPILLAR

PEOPLE WHO LIKE TO GROW THE NEW ZEALAND NATIVE PEPPER SHRUB, KAWAKAWA, WILL BE FAMILIAR WITH THE IRREGULAR HOLES AND NOTCHES THAT MAKE THE KAWAKAWA LOOK SO DISTINCT. THE CATERPILLARS OF THE NATIVE LOOPER MOTH, *CLEORA SCRIPTARIA* — SORRY, IT AIN'T GOT A COMMON NAME — FEAST ON THE LEAVES, MAKING WHAT MOST PEOPLE WILL DESCRIBE AS 'A HELL OF A MESS'. INTERESTINGLY, RECENT RESEARCH INDICATES THAT CATER-PILLAR-DAMAGED KAWAKAWA TENDS TO BE A LOT MORE HEALTHY AND VIGOROUS THAN THE PEPPER SHRUBS WITHOUT THE HOLES.

a moth: a head, a thorax with six legs, and a grossly swollen abdomen is all there is to it; the female has no wings and resembles a simple bag of eggs on six feeble legs.

In order to attract a winged, male moth, she clambers down inside the bag and sticks her long, telescopic abdomen out of the bottom of the case, phero-mones and all . . . After the evening's entertainment, the female will lay her hundred or so eggs inside the case in which she grew up — and then she dies.

When the embryos inside the eggs have developed, tiny caterpillars emerge, which escape from the bottom end of their maternal case into the big, wide

MOTHS AND BUTTERFLIES

world. Some spin a long strand of silk that acts as a rather inefficient parachute, to balloon away from the mother's host tree. This is really the only stage in the life cycle of the bag moth that allows some form of natural dispersal.

Impact

These endemic caterpillars are no more than entomological secateurs — chewers that carry out a bit of pruning (pictured below left). Well-established shrubs and trees should be able to withstand such a natural treatment, simply by growing more foliage or branches to make up for the herbage that gets nicked by the bag moths.

Gardeners may not like the look of holes in palm fronds, or the browning of conifer foliage, but in my view that is merely a matter of aesthetics, not plant health!

Control

If you must, you can use just about any insecticide you like on the poor old bag-moth caterpillar.

Personally, I use digital control as it's far more friendly to the environment in general and to the bag moth in particular. The idea is not only simple, but satisfying as well. Once you get your eye in it is a piece of cake to collect the majority (if not all) of the larval cases from a threatened shrub or tree. Plonk them in a large jar so you can take them to a similar host tree in the nearest park, reserve or piece of bush for release.

Parasites

There is one other good reason to abstain from the use of pesticides when a dense population of *Liothula omnivora* becomes evident. When I tried to rear adult moths from a huge number of cases last year, I was terribly frustrated in my attempts. From the

150 or so specimens I did not rear one single moth, but many hundreds of dark blue, blowfly-like flies — the bag moth's Tachinid parasitoid, *Pales marginata*.

It isn't really surprising that the Tachinid fly has a great nose for its host, and once *Pales* is part of the bag moth's colony, subsequent flies will have no trouble finding more caterpillars to infect. But how does this fly manage to get its eggs and maggots inside the well-protected caterpillar?

The answer lies in careful oviposition. The gravid female parasitic fly lays her eggs on the edge of fresh chewing damage made by a caterpillar. The rationale is that when the bag-moth caterpillar resumes feeding the next night, it over-bites the egg and swallows it whole.

Similarly the fly can lay an egg or two on the top edge of the caterpillar's case, so it will ingest the egg when it's time to make extensions to the sleeping bag. In both methods, the egg gains entry to the host's gut system, where it will hatch into a hungry, meat-eating maggot. The maggot will initially only eat the non-essential tissues inside its caterpillar. If you rely on a supply of fresh, live meat there's no point in killing your host on day one. Best to leave that till the end of the development cycle. By the time the maggots form their brown, barrel-shaped puparia inside the bag moth's case, the caterpillar is a mere dried-out mummy (pictured opposite right).

A few weeks later, the new batch of adult Pales flies hatch from the puparia and exit the case through the narrow bottom end to blow up their two wings and harden up for life. These flies will hang

CLEVER KOKAKO

THERE'S ONLY ONE WAY TO COAX A BAG-MOTH CATERPILLAR FROM ITS COSY HOME, AND THE KOKAKO HAS PERFECTED THE TECHNIQUE. THEY GRAB THE CASE IN THEIR CLAWS BY THE POINTY, BOTTOM END AND FORCE THE CATERPILLAR OUT THE TOP END WITH THEIR GENTLY SNAPPING BEAK — A BIT LIKE PRESSING TOOTHPASTE OUT OF A TUBE!

around the bag moth tree, mate, and continue to increase the infection rates.

Sometimes the cases of bag moths show tiny circular exit holes. These are made by very small parasitoid wasps that complete their life cycle inside the maggots of the parasitic flies. Yes — there are parasitoids living inside the bag-moth caterpillar's parasitoids!

How these minute wasps manage to locate and then infect a maggot inside a caterpillar, which itself lives inside a bulletproof case, is a mystery.

Codling moth

THE CODLING MOTH (*CYDIA pomonella*) belongs to the leafroller Family and is a ubiquitous pest of pipfruit and some stone fruit. Apples and pears are the most obvious hosts, but walnuts, plums,

MOTHS AND BUTTERFLIES

peaches and nectarines are also on the menu. Last century this species was accidentally introduced from Europe. No doubt early settlers took 'maggoty' apples on their journey to New Zealand, and the subsequent generations of codling moths (pictured below) had no problem finding suitable hosts in the developing gardens of the South Pacific colony.

Life cycle

In the winter months the beast over-winters as a full-grown caterpillar in nooks and crannies of the host tree's bark. Deep inside its silken cocoon the caterpillar sits out its diapause — a bit like being in a very cold sin bin for three months — until rising spring temperatures kick the hormonal changes into action. Then by the beginning of October, it's time to moult into the chrysalis stage.

The chrysalis stage, or pupa, is the intermediate phase in which meta-morphosis takes place from a plump larva to a winged and fully mature adult insect. It is also the stage that plays an important role in 'resetting' the biological clock, so that moth emergence coincides with the flowering of host trees in spring.

When it emerges, just about the first thing that's on the blighter's mind is finding a suitable sexual partner. It happens all the time, doesn't it?

The female moth is equipped with pheromone glands, which release minute quantities of a highly attractive scent, sure to draw the attention of any male moth that may sit downwind from her. In fact, the males are preparing themselves for this pheromone, as they spend a lot of time polishing their extremely sensitive antennae. These amazing, almost feathered appendages are able to detect the presence of just a few molecules of pheromone per litre of air.

As soon as the pheromone plume reaches the male, he gets the equivalent of an 'erect antenna' and takes to the air in an upwind direction — Romeo ain't stupid and the rest will be history . . .

The fertilised female will then select some suitable sites to lay her eggs. The small apples (known as 'codlings' in old-fashioned English) that are formed after flowering are the preferred oviposition places, and a week or so later these proto-apples will be the target of the first instar caterpillars that hatch from the eggs.

Impact

The larval form of any insect species is the phase that does all the feeding and growing, and codling moth caterpillars are no exception. They head straight for the fruit and damage the skin by tunnelling into it. Once they're inside you'll have lost the battle: trying to get them out of there is a waste of time and, in addition, the fruitlet

has been opened up for easy entry by fungi and bacteria.

The caterpillars make their way to the core of the fruit, where the seeds or pips are developing. Classic damage consists of a tunnel to the core and copious amounts of frass protruding from the entrance hole. Cavities filled with brown frass and rotting fruit pulp in the centre of the fruit do not taste very nice, as many ardent apple-eaters will have found out at some stage.

Five caterpillar stages are usually enough to complete development, then the full-grown larvae, fat and plump, with a pinkish tinge, leave the fruit and pupate under the bark of the host tree. A second generation of adult codling moths will follow a few weeks later.

Control

Orchardists and hobby fruit-tree growers have cobbled together a cunning chemical spray programme which starts after petal fall and aims to cover the young fruit with residual insecticides in an attempt to kill the susceptible little instar larvae before they enter the fruit. When I searched through my old advisory leaflets I read about a maldison spray at petal fall, followed by carbaryl sprays at three-weekly intervals until the fruit is harvested. Just plaster the fruit and hope for a good kill.

The problem is the codling moth has a varying number of generations in different parts of New Zealand. In Nelson there may be only one generation per year, whereas in Canterbury and the lower North Island a partial second generation

may occur. Hawke's Bay tends to have two generations and the north of the North Island can go to as many as three in one season. This throws the concept of a 'regular' biological clock out the window in a big way.

In the districts where the codling moth has just one generation, the emergence of the adult moths is spread out from late October to late January; no nice single wave of emergence after petal fall. It's as if the species puts its collective eggs in different 'meteorological baskets', so that there will be larvae born in just about every month of spring and early summer.

In areas with two generations of codling moth there may be a sizeable overlap. While caterpillars of the second generation are entering the fruit, larvae of the first batch are exiting. Apart from the fact that this may cause severe headaches for traffic control, it also demolishes the idea of a perfectly timed spray application, based on the 'biological clock'.

Enter the pheromone trap (pictured below). These are simple sticky traps baited with a capsule filled with synthesised female codling moth pheromone (loads of it), and are available from most garden centres and other retailers.

CARDBOARD COCOONS

FRUSTRATED BACKYARD APPLE-GROWERS HAVE OFTEN ADVOCATED TYING A BAND OF CORRUGATED CARDBOARD AROUND THE TRUNK TO PROVIDE FULL-GROWN CATERPILLARS WITH AN EASY COCOONING SITE. THE IDEA IS THEN TO BURN THE CARDBOARD AND ITS INHABITANTS DURING THE WINTER MONTHS, THUS REDUCING THE POPULATION OF OVER-WINTERING LARVAE.

UNFORTUNATELY ALL YOU ARE DOING IS PROVIDING EXTRA COCOONING SITES. THE SPECIMENS THAT ARE BURNED WITH MERCILESS GUSTO WOULDN'T HAVE MADE IT THROUGH THE WINTER ANYWAY, BUT THE REALLY CLEVER ONES REMAIN COSILY HIDDEN UNDER THE BARK IN THEIR SILKEN COCOONS PROTECTED FROM THE RAINY, STORMY AUGUST NIGHTS.

This stuff bombards the poor males' antennae, giving them the impression of a particularly large and receptive Juliet sitting just upwind. When Romeo finally gets to the origin of his desires, he will more than likely find himself stuck with his family benefit on the sticky substrate of the trap. Researchers euphemistically call this 'mating disruption'. A slightly different technique of mating disruption involves the introduction of numerous 'pheromone straws' into the orchard to saturate the air with confusing smells, so very few male moths, if any, succeed in finding a real female.

The guts of the pheromone traps is not so much the fact that you remove a good number of males from the population, but that by regular examination and counting of the trapped specimens you get a pretty good idea of peak flight times. This can then be translated to peak hatching of tiny but very hungry caterpillars a week later. The rest is easy: cover your fruit with suitable insecticides or deterrents and the protection is in place.

These days the traditional organo-phosphates are rapidly falling out of favour so here are some alternatives.

Btk (e.g. Thuricide, Dipel, Agree, Foray) is probably the most target-specific general caterpillar spray (see page 14).

Young codling moth larvae are very susceptible to a 'granulosis virus', marketed as Capex. This virus reputedly shows a better efficacy than Btk at that very young larval stage of the life cycle.

Derris dust is an old-time botanical insecticide or 'organic' compound used with great effect as a protectant against caterpillars. The active ingredient, rotenone, is not necessarily non-toxic to humans and will definitely kill any kind of insect (friend or foe) that may come in contact with the dust. It is also toxic to fish.

The latest botanical insecticide on the market is neem oil. It has a strong repellent and anti-feedant action but is also a general insecticide with no regard for predators and parasites. A new formulation of neem oil (NeemAzal-T/S), extracted by watery distillation from the plant's kernels, is reported as being mortally efficacious on early instar larvae.

In the commercial fruit growing industry a product by the name of Mimic is used to interfere with the moulting process of caterpillars, thereby breaking the life cycle of the insect. Here again, only caterpillars are the target, and feeding stops within hours of ingesting the residue. These so-called insect growth regulators (IGRs) have become a real force in pest control, because they act on very specific hormones within the insect's metabolic pathways.

Predators

Often competition for over-wintering sites leads some Johnny-come-latelies to cocoon on the ground where a good number of predatory insects, such as ground beetles, and birds gobble them up.

Cutworm

CUTWORMS ARE THE CATERPILLARS of large, noctuid moths. They love their plant material young and tender.

Impact

Sometimes whole rows of seedlings are chewed off at the base, leaving the top parts wilting on the ground. The greasy cutworm (*Agrotis ipsilon,* pictured right) is a particular master at this, in early spring. During daylight hours they retreat to a nearby subterranean tunnel — not easy to find.

I don't mind a bit of 'seedling weeding' from time to time, but nice, straight rows of carrots, spinach, radish, beetroot and other quality veggies look rather odd when three or four immediate neighbours have conked out.

Control

Control or prevention of greasy cutworm and related armyworm 'seedling hackers' is not easy, especially if you don't want to chuck persistent insecticides on your tender young veggie plants, but a there's a clever bait mixture that might just do the trick.

The bait uses the caterpillar's attraction to wheat bran. Mix 500 grams of bran with 25 millilitres of vegetable oil so that it becomes nice and sticky. Add 30 grams of permethrin or carbaryl insecticide powder and thoroughly mix to create a lethal greasy-cutworm delicacy that can be sprinkled among the threatened seedlings. You really won't need much to get those rogue chewers.

If you'd rather use something less synthetic, use Btk to make your cutworms crook.

Gum emperor moth

THE GUM EMPEROR MOTH (*Antheraea eucalypti*), not surprisingly, lives on eucalyptus trees and is member of the silk moth Family Saturniidae. It was imported into New Zealand in about 1920 and can be found throughout the North Island and in the north of the South Island.

The moth can have a wing span of up to 140 millimetres. It is fawny brown with pinkish tips on the forewings, some black bands across the hind wings and four distinctive 'eye-spots', one on each wing. The body is furry and the antennae are highly sensitive — especially those of the male, which are used to detect females.

Life cycle

The moths fly from October to well into summer, and are often brought inside by the family cat. The females lay dozens of eggs (pictured opposite left) measuring 2 millimetres in length, often in rows on the leaves of a host plant. They prefer eucalypts, but you will find them on Tristania, pepper trees (*Schinus molle*), liquidambar and even the odd silver birch.

After a few weeks, the little, mostly black caterpillars hatch from the eggs. These have a tendency to feed together on the leaf margins. The caterpillars grow quickly and moulting takes place at regular intervals. With each shedding of the larval skin the caterpillar becomes more and

more colourful. Newly hatched larvae often devour their old skins, so that not a scrap of mineral or vitamin is lost.

When they get to their final instar, the caterpillars are truly spectacular. The body colouration is a vivid blue-green, perfectly matching the gum leaves they feed on. But the ornamentations are something else: brown-purple legs, pale stripes along the mid-line of the body, and yellowy-orange spiky protrusions on the back, which are tipped with the most gorgeous blue and purple (pictured below).

At this stage the caterpillars will be at least 90 millimetres in length. They feed voraciously, literally stripping the leaves until all that's left are the central veins. On the other side of the animal, droppings are produced that are the size of raisins. You'll often find the deep piles of dung before you discover the caterpillars!

Just before pupation the larvae turn a deep purplish colour and the hunt is on for a nice place to spin a cocoon. The cocoon is made of a huge quantity of brown silk. This is not really the stuff fashion designers would get excited about, as it is inferior in quality to the silk produced by the captive silk worms of the Far East.

These oval cocoons are 30–40

millimetres long and are attached to a sturdy trunk, branch or twig, near where the feeding took place. The silk toughens up and is almost impossible to cut open.

Inside this small cocoon (pictured below right) the purple caterpillar takes off its last larval skin and changes into a pupa. There it completes its metamorphosis.

When an adult moth emerges from the pupa it rasps away at the cocoon, with the aid of some sharp hooks on the edge of its folded and crumpled wings, until there is an opening just big enough for the moth to emerge. You can hear it doing this, as it produces a loud rasping noise.

Impact

On well-established trees, the gum emperor moth and its larvae do very little structural damage. In nurseries, however, a few can do quite a bit of damage, as young trees simply don't have that many leaves to spare.

Control

Any insecticide or organic caterpillar killer will do the ghastly deed, if you're sick of the old and laborious method of 'digital control'.

Predators and pathogens

There are some natural predators and even viruses and fungi that kill and gobble up caterpillars.

Wasps (both the German and common wasps, and also the thin paper wasp species) hunt visually and it is not difficult for them to detect their Antheraea meal. Once a group of caterpillars is detected, the wasps will come back to the same branches to harvest as much meat as possible. The wasps simply kill the larvae and cut them up into manageable sections for transport to the nest.

But there are invisible natural 'predators' as well: sometimes caterpillars start to show serious signs of lethargy, and their normally nice, firm droppings become wet, green and stinky. That indicates a virus has hit the population.

But the prettiest pathogen is a fungus called Beauveria. It enters the small to mid-sized stages of caterpillars, completely filling their body with white hyphal material that drops them to the floor within a matter of days.

Monarch butterfly

THE MONARCH BUTTERFLY (*DANAUS plexippus*; pictured below) is one of the few insects that people mention when they are asked to name some beneficial invertebrates. Many people even have the odd swan plant in the backyard for the sole purpose of attracting and breeding these wonderful orange-and-black Lepidoptera.

When you observe a caterpillar carefully, you'll notice that it has three pairs of jointed legs attached to the three thoracic segments, just behind the head, followed by a group of four pairs of abdominal prolegs (false,

unjointed legs) and a pair of anal prolegs right at the end of the body. The prolegs are muscular protrusions that act as legs and help the larva to get a good grip on surfaces. The key to this grip is a row of tiny curved hooks at the end of the prolegs — almost like a row of small crampons.

Telling a male butterfly from a female is child's play. The males have a single dark patch on one of the veins in the hind wings. It's their pheromone pouch, used for mate recognition and courtship.

In autumn the butterflies disperse and will over-winter in groups, hanging from trees that afford them a foothold and shelter from the cold winter winds.

There are some famous over-wintering sites where thousands of butterflies get together to keep each other company during the cold winter months. Butterfly

GUM CHEWERS

EUCALYPTS ARE A FASCINATING GENUS. THEIR LEAVES ARE FULL OF STRANGE, AROMATIC AND VOLATILE SUBSTANCES (TANNINS AND PHENOLICS), WHICH WERE, NO DOUBT, DEVELOPED OVER AEONS TO WARD OFF BROWSING CREATURES. BUT NATURE ALWAYS FINDS A WAY TO OVERCOME THESE KINDS OF BARRIERS, AND I'M PLEASED TO SEE THAT THE ODD MARSUPIAL, PLUS A WHOLE RANGE OF INSECTS HAVE MANAGED TO CIRCUMVENT THE POISONOUS DEFENCES OF THESE TREES. A GOOD RANGE OF INSECTS ARE FOUND FEEDING ON THE EUCALYPTS AND, OVER THE PAST HUNDRED YEARS OR SO, A NUMBER OF THESE AUSSIE GUM CHEWERS HAVE MADE IT TO NEW ZEALAND.

Bay (Tauranga Bay) near Whangaroa in
Northland is such a site, as is Church Hill
in Nelson. The adult butterflies rest in
dense trees when temperatures are below
10°C, but on warmer winter days they
may go for short feeding flights and return
again in the afternoon.

Life cycle

The monarch's life cycle is one of
complete metamorphosis, involving the
four main stages of development: egg,
larva (caterpillar), pupa (chrysalis) and
adult (butterfly). The tiny caterpillar that
emerges from the egg will often consume
its eggshell as its first feed, before tackling
the green leaves of the host plant.

Larvae are the feeding machines
responsible for the actual growth of the
insect. There are five caterpillar stages
(called instars) in the life cycle, each
preceded by a skin moult.

The last larval skin moult, from final
instar caterpillar to pupa, is the tricky one.
A full-grown, fat, last instar caterpillar will

select a suitable site for pupation: underneath a branch, windowsill, horizontal
piece of timber on a fence, etc. They
don't always pupate on the host plant
and may wander quite a distance. At
the chosen spot the caterpillar uses its silk
glands to spin a small silken pad onto the
support surface. Both the caterpillar and
the pupa will use this pad as a base to
hang from during the transformation.

The full-grown caterpillar hangs
from the silken pad (pictured left) with
the crampons of its anal prolegs, and
assumes the characteristic J-form for
a day or so. During this 'pre-pupal'
stage, the colours of the caterpillar fade
because of hormonal changes inside the
body.

The larval skin splits open along the
back, and the green pupa, adorned with
golden spots, becomes visible. The insect
then moves and wriggles to free itself
from the old wrinkled caterpillar skin,
which is slowly moved upwards in the
direction of the silken pad.

But the insect is still attached by its
larval hind legs to the pad, while the
larval skin needs to be discarded! It's time
for the pupa's special trick.

The newly formed pupa possesses
a black stalk, armed with Velcro-like
spines, at the end of its abdomen
(pictured overleaf). The idea is to attach
that stalk (or cremaster) to the silken
pad before the caterpillar skin gets
discarded completely. Thus the pupa
can hang all by itself and complete the
metamorphosis.

The pupa performs this death-defying act with the help of two opposable and movable knobs on the belly side of its skin, just down from the cremaster. With these knobs the pupa grabs a firm hold of the old skin, allowing it to free the cremaster and 'jab' the Velcro hooks into the silken pad. It can take a while to find a foothold, but eventually the pupa will hang from its cremaster. More vigorous wriggling will ensure that the old skin is dislodged from the pad.

Sometimes things go horribly wrong here, causing the pupa to crash to the ground, often with tragic and messy consequences, but even when the pupa survives the fall it will not complete the transformation into a monarch butterfly unless it's hanging upside down. When you find a fallen pupa on the ground and it still looks OK, the idea is to simply hang it back up via its cremaster — if you cannot find a silken pad, make an artificial one from . . . yes . . . Velcro!

Food sources

The most common plant we use as food for the caterpillars is the good old swan plant (*Gomphocarpus fruticosus*). It belongs to the plant Family Asclepiadaceae, or milkweeds. When you break off a leaf or twig, you'll find out why it is called milkweed, and when you taste the milky-white sap you'll find out that was a pretty stupid thing to do — it contains caustic cardiac glycosides. This is the reason some childcare centres are wary of growing swan plants.

Monarch butterfly caterpillars and some of their relatives have adapted to the consumption of these leaves and their toxic contents. What's more, they use the cardiac glycosides as personal protection against vertebrate predators, and carry it with them through into the butterfly stage. No wonder that all life-cycle stages can afford to be conspicuous and colourful — the colours act as a warning: 'One bite, my friend, and you're going to end up with a great gut-ache!' Indeed, you'll rarely see a bird catching a monarch butterfly or devouring a caterpillar, so the warning signs seem to work quite well.

There are a number of very good alternatives to the swan plant available in New Zealand. First of all there's the giant swan plant, *Gomphocarpus (Asclepias) physocarpa*. It tends to be a larger plant with larger seed pods, but still with the same boring creamy-white flowers.

The bloodflower (*Asclepias curassavica*), so called because it bears nice umbels of orange-red flowers, has a bushy habit and won't grow much taller than a metre. The

eaves are green, tinged with red, and plants tend to regenerate quickly after caterpillar defoliation. It is not terribly frost hardy, but I'm still amazed at how well it performs here in New Zealand, after seeing it 'in the wild' in the tropical coastal region of Ecuador and in the Amazon river basin. It is a prolific seed producer, so even in colder parts this food plant can be grown year after year with a bit of careful management.

A close relative is the butterfly bush (*Asclepias tuberosa*) from North America. A similar height to the bloodflower, it has orange and yellow flowers, but is more cold-tolerant and a perennial. It has a tuberous root system, enabling survival during cold winters — down to −20°C! This plant could well be the answer for those people wishing to maintain monarch populations in Canterbury and Otago.

Asclepias 'Silky Gold', also from North America, produces yellow flowers on a bush which grows to 1.2 metres tall. Broader yellow leaves characterise this accepted food plant species.

So there you are: a whole new range of food plants for monarch caterpillars. The female monarchs will sniff them out and lay eggs on these hosts without any problem. Most of these milkweeds will, from time to time, get a colony of yellow oleander aphids around the growing tips. Simple 'digital control' or an application of potassium salts (fatty acids) will control these suckers, and won't do the caterpillars or pupae of the monarch any harm at all.

The most common problem over the summer months is running out of food for the caterpillars. It's the old trap for young players in the monarch game: the butterflies keep on laying eggs on an already diminishing host plant.

Control

Four to five caterpillars on a 1 metre high swan plant is ample: it allows the plant to regenerate while some selective harvesting is going on by the residents.

Newly laid eggs are relatively easy to spot and should be squashed there and then. It sounds awful, but better to do it at the egg stage than having to see your half-grown, starving larvae wander off, feebly crawl over the top edge of the fence and exhaustedly fall to their death on the concrete path, don't you think?

An alternative strategy is to maintain plants with the maximum allowable number of caterpillars under a tent constructed from bamboo and curtain netting, keeping them out of reach of ovipositing female monarch butterflies as well as predators such as wasps. You can always keep spare, clean plants in a shade house or glasshouse on 'stand-by'.

If you do get into a desperate situation and have no spare plants, there are a few food items from which the caterpillars can obtain some sustenance, although the monarchs will not lay their eggs on them. Tweedia (*Oxypetalum caeruleum*) is a low-growing perennial with greyish leaves and pretty blue flowers. When damaged, Tweedia bleeds a white, milky sap. Caterpillars can and will eat the new leaves of this plant.

PACIFIC SPREAD

THE MONARCH BUTTERFLY IS OF AMERICAN ORIGIN AND GOT HERE ALL BY ITSELF. ENTOMOLOGISTS ARE UNSURE AS TO WHEN THE MONARCH BUTTERFLY REACHED NEW ZEALAND; THE BEST SCIENTIFIC GUESS IS 'BETWEEN 1841 AND 1873'.

WHEN WE TAKE A LOOK AT MONARCH DISPERSAL THROUGH THE PACIFIC IN THE 19TH CENTURY IT BECOMES CLEAR THAT THEIR ARRIVAL IN NEW ZEALAND WAS INEVITABLE. THE SPECIES REACHED HAWAII IN 1840 AND TONGA IN 1863. RAROTONGA RECORDED ITS FIRST BUTTERFLY SIGHTING IN 1869, AND AUSTRALIA FOLLOWED IN 1871. THE INTRIGUING THING IS THAT BEFORE 1840 THE MONARCH WAS NOT KNOWN TO OCCUR OUTSIDE THE AMERICAN CONTINENT AT ALL. WHY THE SUDDEN MOVE IN OUR DIRECTION?

THE ANSWER MAY WELL BE THAT THE MID-1800S MIGRATION WESTWARDS WAS NOT THE FIRST DISPERSAL ATTEMPT, BUT THE FIRST SUCCESSFUL ONE. YOU CAN FLY WEST ALL YOU LIKE, BUT WHEN IT COMES TO SECURING A FUTURE FOR YOUR OFFSPRING, AND HENCE YOUR SPECIES, YOU NEED A SUITABLE FOOD PLANT. THERE IS STRONG EVIDENCE THAT MILKWEED (*ASCLEPIAS* SPECIES) WAS DISTRIBUTED THROUGHOUT THE PACIFIC BY SAILORS WHO USED THE SOFT, FLUFFY SEEDS OF THIS PLANT AS STUFFING FOR THEIR BEDDING.

Some folks rely on one of the most horrible weeds in the northern part of our country — moth plant (*Araujia sericifera*), also known as moth-catching plant, cruel vine or kapok vine. It is an invasive climber that smothers trees, shrubs and bush, and you can often find it climbing all over fences in deserted gardens and sections, even right in the middle of the city of Auckland. This plant must be dealt with severely, and should ideally be exterminated from all gardens in New Zealand.

Hungry monarch caterpillars will also complete their life cycle on the flesh of cut-open pumpkin or the skin and flesh of sliced marrow fruit and telegraph cucumber. But only the largest caterpillars (the ones that have fed for most of their lives on the proper Asclepiadaceous plants) will successfully make it to the pupal stages, without turning out as deformed butterflies.

Towards late autumn the incidence of crippled butterflies increases dramatically, and this is no doubt caused by the plummeting ambient temperatures. A larva, but especially a pupa, needs a minimum amount of 'degree-days' for proper development. It is therefore not a good idea to persevere with new eggs and larvae in autumn. What is the cut-off date? I always say 23 April because that is an easy-to-remember date for my mother and me, but Anzac Day will do.

Predators

Paper wasps (Polistes species) and European wasps (Vespula species) are basically protein-feeders and hunt by visual cues.

Monarch caterpillars with black and yellow striped pyjamas are an easy-to-find prey, and the wasps will cut the larvae to bits and carry them back to the nest. They'll even find and remove the eggs, presumably to construct an omelette at home. Pupae are often overlooked by these introduced pests.

If you can find the wasps' nests and (carefully) destroy them, it will certainly help to save your monarch butterfly farm. As indicated before, the plants plus their valuable cargo can be protected from the wasps by covering them with curtain netting or shade cloth, draped over a timber or bamboo framework. Another good idea is to keep your food plants in large pots which can be moved, preferably to a shade house or glasshouse for protection.

It may come as a shock to you, but praying mantids need to live too, and monarch caterpillars are as much on the menu as flies or moths. The remedy is simple: take the mantids off the monarch's food plants and relocate them so that they can do a lot more useful preying, such as in the vegetable garden.

During the warmer days of the monarch season, caterpillars can suddenly display puzzling symptoms: they hang from the twigs as limp, dark skins, oozing a yukky dark green to black soup. This could be caused by a viral infection, especially when plants are overcrowded. Infection can spread quickly and quarantine may prevent mass destruction.

Similar symptoms can also be caused by a rather elusive, native predator, the soldier bug, from the same insect family as our green vegetable bug. It looks like a green veggie bug, but it is brown and slightly smaller. Soldier bugs use their proboscis to suck fluids out of their caterpillar prey, performing a population-regulation function. Have a look for them when you see those limp caterpillars draped over the twigs of the swan plant — they're fascinating predators, no matter whose side you're on!

Painted apple moth

ONE OF THE LATEST insects to get a foothold in New Zealand is the painted apple moth (*Teia anartoides*).

Teia anartoides is a moth belonging to the Lymantriidae Family. Some of the members of this Family are serious forestry pests. Think of the gypsy moth and the old white-spotted tussock moth, which was eradicated from east Auckland's suburbs by an aerial spraying programme — Operation Evergreen — in the 1990s.

The beast is native to Australia and was first reported in the Auckland suburb of Glendene early in May 1999, followed by sightings in Mount Wellington towards the end of September of that same year. In late February 2002, planes and helicopters began to apply Btk over the target zones in west Auckland, amidst public protests. To say that the painted apple moth eradication project created a storm, socially and politically, is an understatement.

The male's forewings show various

MOTHS AND BUTTERFLIES

shades of brown and black in a pretty pattern, adorned with white markings. The hindwings have a large patch of orangey-yellow with a black border. When sitting at rest, the male is about 13 millimetres long, with a 25 millimetre wingspan.

The large, feathered antennae on the head and the orange spots on the furry thorax and along the legs add to this moth's exotic appearance. Often brightly coloured moths act like honorary butter-flies and are active during the day, rather than at night. The painted apple moth male is no exception. It flies in daylight hours in search of a receptive female.

The females are comparatively boring brown and hairy insects, with minute stumps of degenerate wings that are no longer any use at all. These female moths are merely living bags of eggs on six legs — a Juliet waiting for a Romeo to fly by.

Mind you, these females make clever use of a complex set of sex-attractant chemicals called pheromones to lure their prospective mates into a close encounter.

Life cycle

After mating, the female will lay hundreds of off-white eggs around her thinning body, often creating a soft 'nest' from the leftovers of her pupal cocoon and the hairs and scales that are shed from her body.

The eggs hatch into small orange-brown and black caterpillars (pictured right) — little, hairy, voracious chewers that waste no time getting stuck into the leaves of their host plants. In the quiet west Auckland suburbs the larvae are mostly found on their preferred hosts — wattle species. Ironically, these trees tend to be viewed as invasive weeds.

The caterpillars grow with every skin moult and reach a length of about 35 millimetres when fully grown. At this stage, they are the prettiest critter imaginable, with their orange-yellow body, red-brown head and legs, and black striped back.

The whole insect is rather hairy, but behind the head two long, black tufts of hair stand straight up into the air. On the back of the body, four distinct brushes of dense white or grey hairs complete the exotic look of the painted apple moth caterpillar.

The life cycle is completed when the caterpillars spin a flimsy, silken cocoon (pictured opposite left) on or near the host plant, inside which they change their skin once more to become a pupa or chrysalis. Pupae can be found away from the host plant, on fences, houses, outdoor furniture, containers etc. This, incidentally, may indicate a possible means of transport from Australia to New Zealand.

There are some remarkable ento-mological aspects to the behaviour and life cycle of this insect that not only make it

an interesting creature to study, but also difficult to eradicate from infested areas.

First of all, the female does not fly at all. At first sight that could be perceived as an advantage, as the species is not likely to spread too fast. At least that's what MAF thought: 'a wee bit of time to do some research and draft a plan of attack.' But the painted apple moth had a surprise in store.

Even without the help of humans, the painted apple moth appears to be perfectly capable of spreading at a rate of knots. The newly hatched, young caterpillar is known to be the 'flying stage' of the life cycle.

They crawl to the tops of trees and reel out a long, thin thread of silk that is picked up by a gentle breeze. All these small, light caterpillars have to do is let go of their host tree and hang on to their silken 'balloon'. The wind will take them a long way downstream, so to speak, where they may find another host tree to feed on.

If you feed on a wide range of host trees, your chances of finding a suitable host increase dramatically. The viability of a single male or female emerging from the pupa in a brand-new district can be regarded as low, since not many of your compatriots may have made exactly the same journey through the air on a piece of flimsy silk. But the painted apple moth's powerful pheromones come in handy here!

The male moth (pictured below right) is equipped with large, feathered antennae that catch the merest whiff of female pheromone and cause an instant reaction, and a searching flight in the direction of the 'calling' female.

Impact

The Ministry of Agriculture and Forestry has conservatively estimated the potential cost of painted apple moth, if established in New Zealand, at $48 million for the next 20 years, and that is excluding the impacts on horticulture, trade access and the conservation estate.

Painted apple moth caterpillars have been recorded on a wide range of plant species in Australia — more than 50 types of plants in 24 Families. It's the same here, where they go for commercial forestry species as well as amenity plants and natives — the list of confirmed hosts continues to grow.

MOTHS AND BUTTERFLIES

Trapping

In the horticultural industry, pheromones of pest insects are used to detect the presence of a species, or to warn growers of a population explosion, based on trap catches. In the case of the painted apple moth it is also important to know exactly where your target is before you can attempt eradication, so the basic idea to use the moth's own, unique lure in monitoring traps was a good one.

The problem was that the pheromone turned out to be a complex mixture of compounds that was not easily identified, let alone synthesised for use in traps. Two competing teams of capable pheromone scientists had a go at cracking the moth's code, initially with little success.

In the absence of a good, synthetic pheromone, the next best thing was to bait the traps with live, caged females, which were checked on a regular basis. About 700 traps showed the distribution and expansion of the population in west Auckland. They also showed patterns of more-or-less synchronised life cycles and that development continued during the colder winter months, albeit a lot slower than in summer.

There were people critical of the fact that the moth was not 'dealt to' within the first few weeks of detection. There is no doubt that with the benefit of hindsight an operational plan for eradication can always be improved.

But what would have happened if MAF started squirting chemical or biological insecticides from choppers in week one? The critics would have pointed out that such an action — without some preliminary research, official permits or consents and without adequately informing the public — was unacceptable.

And how embarrassing would it have been had MAF gone off half-cocked with

THE BENEFITS OF ISOLATION

QUARANTINE AND BIOSECURITY RECEIVE A LOT OF MEDIA ATTENTION. THE ASTONISHING REGULARITY WITH WHICH NEW PESTS AND DISEASES MAKE THEIR WAY INTO NEW ZEALAND IS OF CONCERN. OUR UNIQUE, ISOLATED POSITION AT THE BOTTOM OF THE GLOBE HAS BEEN ONE OF OUR GREATEST ASSETS AND WE HAVE WONDERFUL FLORA AND FAUNA THAT WERE CREATED BECAUSE OF THIS ISOLATION. MOST OF US REALISE THE SEPARATION FROM THE OTHER PARTS OF GONDWANALAND, SOME 80 MILLION YEARS AGO, EFFECTIVELY GAVE US THE GREAT ADVANTAGE WE STILL ENJOY — WE HAVE RELATIVELY FEW PESTS AND DISEASES COMPARED TO OTHER COUNTRIES THAT PRODUCE MEAT AND DAIRY PRODUCTS, TIMBER AND FOOD CROPS. THAT GIVES US A REAL ECONOMIC ADVANTAGE. BUT, EVEN IF WE LOOK BEYOND THE SIMPLE ECONOMICS OF OUR QUARANTINE STATUS, OUR NATURAL BIOTA ARE UNIQUE, DESPITE THE FAMOUS INTERLOPERS — WEEDS AND PESTS — THAT HAVE MANAGED TO SPOIL THE ECOLOGICAL FUN.

spraying programme in west Auckland, only to find that the species was also established in other areas? Delimiting surveys are a fairly standard set of procedures to convince officials that they are not wasting their money.

The trapping programme was devised to get an idea of the moth's spread, but interpretation of the catches relies heavily on our knowledge of the beast's behaviour.

How far do lone male moths fly in search of a female — hundreds of metres, or kilometres? The mark-release-recapture trials that were carried out indicated some strong flying.

But even the knowledge of how far a winged male moth can fly does not give an indication of the distribution distances of ballooning larvae. Invertebrates that passively travel through the air can move many, many kilometres — some are known to cross oceans!

Another problem was the matter of funding. Government doesn't always approve funding for the eradication of a biosecurity threat in a hell of a hurry, and recommended control procedures may become less effective over the period of time it takes to get the budget approved.

Control

When it comes to controlling insects from the air — be it by DC-6 or helicopter — there will always be a problem with the creepy feeling that those chemicals are going to kill everything in their path. I suppose it's the most powerful legacy of Agent Orange and associated war movies.

Indeed, an aerial spray of DDT would not be great idea, nor would a lot of people agree with organophosphates or even synthetic pyrethroids being applied in this manner. Luckily, these days there are some gentler materials that can be used, and they are better targeted too!

Bacillus thuringiensis var. *kurstaki* (Btk) is, after all, a bacterium that is naturally found in soils — even in New Zealand. This organism can be sprayed over foliage and, after ingestion, it is known to affect the gut lining of lepidoptera larvae, causing them to stop feeding and die. Non-target species are, by and large, not affected.

It has been used as an aerial spray in North American forests since 1958 and found to be pretty safe to humans, too — a perfect example of this century's biological pest control.

Towards the end of the campaign, another technique was employed in the fight against this moth — thousands and thousands of sterile male painted apple moths were released in the wild to hunt down the remaining fertile females. The idea is very simple — when a sterile male mates with a fertile female she thinks 'all is well' and lays her eggs which, of course, are sterile and will not produce viable offspring. It is called Sterile Insect Technique (SIT) and is used on many other pest species in the world.

As I write this, in early summer 2004, it looks as if the last painted apple moth has been killed — another year's trapping will hopefully confirm this but so far it looks extremely promising.

Garden goodies

Cockroach

WINTER CAN BE A nightmare for phytophagous (plant-feeding) species, especially when food plants lose their leaves. Little wonder that this season is the natural time-out period for most inverte-brates. But there are some hardy insects that I look out for in winter because they are efficient and beautiful — cockroaches.

Roaches are adapted to survive in the harshest conditions. They can go without food for a good many days, but more impressive is the fact that they can go without water for long periods of time, too. To top it all off, they seem to have a knack of being able to survive in areas with a very low relative humidity.

Life cycle

What really strikes me as awesome is the cockroach's ability to breed quickly and efficiently — and that the females display some endearing maternal traits towards their embryonic offspring.

The eggs are not laid singly, but a couple of dozen are kept together in an ootheca or egg raft, a purse-like brown structure that protrudes from the female's abdomen for many weeks (pictured below left). While the eggs develop, the whole raft is kept moist by the female's body, so that the eggs don't dry out. A few days before hatching, the ootheca will be deposited in a sheltered spot; if the female releases the egg raft too soon, the embryos die. The hatching larvae or nymphs then

ave to feed and fend for themselves.

One of the great roach myths is that if ou squash an adult insect, it will quickly y its eggs and hundreds of babies will be orn in no time — this is simply not true.

Benefits

is crazy that we spend an awful amount f collective money trying to control these nsects. They turn dead, organic material nto useful, small bits that can be further roken down into plant food. It's the ycle that keeps the world ticking over nd our hopes of survival on this planet live, yet they are perhaps one of the most ated invertebrates in peri-domestic tuations worldwide.

In any case, this species has made its ome wherever there are human opulations — if you are one of the vorld's most successful recyclers, you night as well associate yourself with the nammal that produces the largest amount f waste!

So when you see a roach this winter, ave another look at it and admire its retty colours or the intricate pattern on s flat wings. Be aware that the antennae arry thousands of sensors and that under nat almost translucent thoracic shield is a lever little head with some of the oldest nd best composting units in the business — cockroach mandibles.

Impact

'm sure these insects gratefully devour all our spilled spaghetti bolognese and

splashes of orange juice, with their 300-million-year-old motto: 'Don't you worry about a thing. We'll clean this up for you real quick!' The only problem is that they don't often wash their six little hands and can therefore spread all sorts of organisms and diseases from contaminated areas to your food.

Opportunistic — that's a good word for them! They know where to live and that there is safety in numbers when you live with a grumpy mammal. Coupled with that, peri-domestic roaches often display a characteristic quick-retreat-and-hide behaviour when they are disturbed at night by a sudden illumination of their habitat. It's all part of their stunning evolutionary success.

ANCIENT INSECTS

COCKROACHES ARE ANCIENT THINGS — FOSSILS FROM THE UPPER CARBONIFEROUS ERA, ABOUT 300 MILLION YEARS AGO, SHOW THAT ROACHES LOOKED THE SAME THEN AS THEY DO NOW. THEY WERE DESIGNER INSECTS RIGHT FROM THE WORD 'GO' AND HAVE NEEDED VIRTUALLY NO ADJUSTMENT OR ADAPTATION. THERE'S EVIDENCE THAT THESE HANDY BEASTS WERE DOING THE SAME JOB, FOR ALL THAT TIME. I THINK THESE RECYCLERS DESERVE A MEDAL FOR LOYALTY AND PERSEVERENCE.

GARDEN GOODIES

Predators

Kiwis — the birds, that is — love this insect (especially the black stink roach, pictured below, as it is easily detected when you are virtually blind and have to locate your food by touch and smell). I suppose this highlights the importance of all invertebrates in their native ecosystems — they all have a job description and they all are food for another member of their ecological community.

Native cockroach

SOME OF OUR MOST common native roaches may look a little bit like the German variety. They're the same size, up to 15 millimetres long, and brownish and flat — but you can tell them apart quite easily once you know what to look for.

Blattella germanica, the species that gives all other cockroaches a bad name, always has two distinct parallel stripes on the thoracic shield at the front of the insect (see page 144) — our native roaches never have this feature. Our natives occur in ones and twos in the house, and will sit on the wallpaper for hours and hours, even during the day, without being unduly disturbed by approaching home owners. They can see you with their keen eyesight, they can smell you with their sensitive antennae, they can feel your footsteps from miles away — yet they don't show the slightest sign of panic.

You'll never find our native *Parellipsidion* or *Celatoblatta* under the stove, behind the fridge or in the pantry, chewing through the packaging to get at the food, but you will find them coming out of a bouquet of camellia flowers or a basket of firewood, taken from the stack under the lean-to. Decaying organic material in its purest form — bark, lichen, musty leaf litter and a bit of compost, all mixed together — that's their preferred habitat. They also like to sit in the shelter at the base of cabbage tree leaves, or under the flaps of teatree bark — the human thing is not for them. The environment needs to be reasonably moist, otherwise our delicate native roaches dry out and that can be fatal.

One native species, *Platyzosteria novaeseelandiae,* is a bit bulkier than the brown roaches described above. Over 20 millimetres long, with a stout, black body,

his roach has the ability to emit a foul-
melling odour when it is disturbed on the
orest floor. That's why it was named the
black stink roach.

Earwig

MOST PEOPLE KNOW WHAT earwigs
ook like. They are common insects around
he house and garden and there are some-
hing like two dozen different species in
New Zealand. They vary in size and can be
ight to very dark brown or almost black in
colour.

Earwigs have always had a dodgy
press. I am not sure what it is about them
hat causes so much suspicion in the
human race.

Perhaps it is because of their fierce
ooking pincers (or forceps) at the end of
heir abdomen. This is exactly the reason
hese insects are equipped with such tools:
o impress and to scare off competitors and
ven predators. Although they may give us
wee nip, earwigs generally won't be able
o do too much damage to our sturdy skin
— so it's all bluff!

The common name 'earwig'
probably derives from the shape of the
insect's wings. Yes, if you can find the
insect's membranous wing located under
the sturdy, short and thickened forewing
covers and spread it out, it has the distinct
outline of an ear!

So, despite their confusing common
name, Dermaptera have very little to do
with our ears and I'm sure they'd be
rather keen to avoid such a smelly and
waxy habitat.

Benefits

The earwig was designed to convert
rotting, organic materials into nitrogen,
phosphorous and potassium, plus trace
elements, and they do this for free.

In Western Europe, apple growers
actually encourage earwigs into their
orchards to help control aphids.
Orchardists place upturned clay plant pots
filled with straw into the crooks of
branches (pictured overleaf). These pots
attract the European earwig (*Forficula
auricularia*), which also occurs in New
Zealand. The jury — which had been out
for a good number of years — has come
back with the verdict that these humble

insects do far more good than evil, especially in orchards. Their mixed diet consists mostly of aphids and other small sap-sucking insects that damage the crops, so it is worthwhile providing earwigs with nifty little apartments that are recyclable and biodegradable.

Impact

Most folk who have an apple tree in their backyard, when they find an earwig inside an apple, automatically assume that the tunnel was constructed by the earwig. More often than not, the caterpillar of a codling moth or another leafroller species has caused the damage earlier in the year. The earwig merely moves into the second-hand dwelling lined with rotting apple pulp, which is a real delicacy for an insect that makes a habit out of eating decaying organic materials.

However, for the dahlia growers and stone fruit producers, I have to make mention of the fact that earwigs can sometimes be really naughty — during periods when moisture is scarce they tend to make their own organic material by nibbling at tender and moist flower petals, or by eating holes in the skins of juicy stone fruit. When you're thirsty you might as well access the nicest juice you can get, don't you think? Strawberries and some vegetable crops receive similar attention from earwigs.

Control

Is the late summer and autumn damage to your flowers so bad that action is required? Do you need the blooms to be perfect? Is your crop damage causing a loss of income? The economic reality of control lies in the cost of the control measure chosen, and the reduction of damage achieved.

Some people swear by providing earwigs with a shelter in the form of crumpled-up newspaper in an upturned flowerpot. The idea is that the earwigs will crawl into this shelter site, then the newspaper is collected and burned — you literally incinerate the poor blighters.

There's nothing wrong with this technique, but you'll have to remember that earwigs do like to shelter in moist debris and over-winter as adults or eggs in soil and compost. Therefore, when you offer a dry-ish site in the form of newspaper, it may pay to remove alternative and

preferred shelter sites in the form of rubbish, sacks, boxes etc.

There is the clever trick of baiting insects with something they really like. For earwigs, you can use a mixture of 200 grams of wheat bran, 100 grams of blood-and-bone, and 100 grams of sugar, but you'll have to add 30 millilitres of maldison liquid (Mavrik or Target) to make it lethal. This is not an 'organic' solution, but with a bit of common sense you could place the baits in such a way that non-target species will not have ready access to the stuff — try under clay plant pots, inside closed and firmly anchored yoghurt containers with small holes drilled into the sides, or under rocks and other out-of-the-way places birds and puppies can't reach.

Predators

A number of ground-feeding birds will gobble up earwigs: blackbirds, thrushes and chooks to name but a few. Frequent raking of the compost on top of soil will disturb the insects and make it easier for those birds to find them. Other earwig enemies are the predatory invertebrates that cruise the compost and layers of soil: ground beetles, rove beetles, and even larval tiger beetles will capture some nymphs. And then there are the centipedes of every description and size.

Being an earwig isn't easy and with the numbers of predators about, it is amazing that we see these critters in our gardens and orchards at all.

IN YOUR EARS?

A LOT OF PEOPLE WORRY ABOUT EARWIGS AS THEY HAVE A REPUTATION FOR INTERFERING WITH HUMAN BODIES. INDEED, THEY LOVE TO CRAWL INTO CONFINED SPACES TO HIDE FROM DAYLIGHT AND SUNSHINE. TO BE QUITE HONEST, I THINK THEY ARE ACTUALLY MORE INTERESTED IN HIDING FROM DIURNAL PREDATORS, BUT WHO CARES; AS LONG AS THEY DON'T CRAWL INTO OUR EARS, EH?

APPARENTLY, THE NUMBER OF EARWIGS HAVING TO BE MEDICALLY EXTRACTED FROM EARS IS A LOT LOWER THAN FOR ADVENTUROUS COCKROACHES.

Praying mantid

THE BEAST WE CALL 'praying mantis' belongs to the Insect Order of 'mantids' (Mantodea). Therefore a praying mantis is entogrammatologically known as a mantid species.

There are two main mantid species in New Zealand, the native New Zealand mantis and the introduced springbok mantis.

Life cycle

Egg cases, also known by the wonderful entomological term of oothecae, tend to be laid in late summer and autumn, no doubt with the sensible ecological idea that you might as well over-winter as an egg. The *Orthodera novaezealandiae* egg cases (pictured below right) look like miniature versions of Ayers Rock, are mid-brown in colour, and usually contain a couple of dozen eggs.

The springbok mantis makes a fawn-coloured, meringue-type contraption with a distinct 'pan handle' at the end. These cases are known to have up to a few hundred eggs embedded in them, and can be found in the most unlikely places: on window ledges, brick walls, outdoor furniture, under the eaves, and even on the ceiling inside. More natural oviposition places outside include branches and tree trunks.

The young nymphs hatch in spring and develop during the warmer weather. They grow by shedding their skin at regular intervals. There can be as many as six larval stages in a life cycle. The larger nymphs show the wing pads emerging from the thoracic segments. The final skin moult turns the adolescent nymph into a winged adult and that's when all the fun begins . . .

If you have a cyclopean ear like the mantis does, you should be able to hear the moans and groans of the male mantis for miles . . . until his head gets bitten off by his darling partner. This is the famous act of sexual cannibalism. It is not very common in our native mantis, but a regular feature for a whole heap of other mantid species, including the springbok mantis.

Even while his head and thorax are being consumed, he keeps his 'mind' on the job — a feat which is made possible because of neurological centres in each body segment, which can keep supplying stimuli independently from the main brain.

So, why did these sacrificial males evolve? Isn't natural selection all about survival? No, natural selection is all about passing on as much of your genetic material to the next generation as possible, and survival (plus the chance of multiple copulations) is merely one of the strategies that can be employed.

In the case of male mantids it is not always easy to find a mate, because population densities are not very high. So if

ou do find a mate, you might as well make
ure that she makes the best use of your
genetic package. A well-fed female lays
more eggs than a starving one, and protein is
what she needs to produce eggs. By offering
his body, the male mantid ensures a larger
number of offspring with his genes.

If you keep a gravid female springbok
mantis in captivity over winter, you'll find
out how it all translates into reality.
Females can lay up to six oothecae in the
period from April to November. In total,
hundreds and hundreds of nymphs are
liberated in the garden, and this whole lot
is likely to have come from just the one
father, who has been, by now, well and
truly recycled.

Benefits

As soon as the tiny but very active nymphs
hatch from their eggs, they are on the hunt
for meat; not dead stuff, but real live prey.
Initially they'll capture vinegar flies and
other small fry that venture too close to
the 'preying' mantis babies. These
youngsters can be found grazing on an
aphid colony or catching ants and small
caterpillars; but don't be surprised to see
them hoeing into their own siblings as
well, for cannibalism is a sinister aspect of
these cute and popular predators.

As the nymphs grow larger, the
selected prey items get larger too. Often
the gluttons grab a piece of prey that
appears to be too large. Seeing a medium-
sized nymph trying to manipulate a large
blowfly, with frantically beating wings, is a
real hoot (pictured on page 144): the

mantis must get numb legs and brain cells
as it gets violently shaken about. But it
keeps on hanging on, and will even apply
the mandibles on this struggling prey!

New Zealand praying mantis

NEW ZEALAND'S 'NATIVE' MANTID
species used to be known as *Orthodera
ministralis* (pictured overleaf) just like the
Australian species but is now called
Orthodera novaezealandiae (meaning 'from
New Zealand'). No doubt our mantid
came from Australia, but whether or not
it was a human-assisted (accidental)
introduction in recent times is still being
debated. (This has great bearing on the
validity of the word 'native'.)

The adults of the native species are
winged and the wings fully cover the
abdomen in both males and females.
What's more, both sexes can fly.

Springbok mantis

IN THE 1970S, A second species of
mantid became established. Originally
from South Africa, it goes by the
common name of springbok mantis. This
beast, *Miomantis caffra,* was first found in
the Auckland suburb of New Lynn and is
slowly spreading through the country. It
occurs as far north as Kaitaia and as far

south as Hawke's Bay, but it is still on the move.

The springbok mantis female cannot fly. Her wings do not cover the whole length of her abdomen. This is especially noticeable when she is very pregnant. Springbok mantis males do have a full wing cover over their abdomen and they are able to flutter about quite happily, although not very gracefully.

An interesting oddity of the springbok mantis is that, although the majority of the population is green, some specimens can be fawn, brown, or even yellowish-pink in colour. These are genetic variants.

To the untrained eye, both mantid species look rather similar, but the differences are easy to spot if you know what you are looking for.

If you think you've identified your adult, winged specimen, you can always check their raptorial legs: the inner side of the native mantis's front legs is yellow with a vivid dark blue patch on it (pictured below); the adult springbok mantis' front legs are not similarly adorned.

Another distinguishing feature is that the thoracic segment, just behind the head, is much broader in the native species than in the South African newcomer.

There appears to be a bit of a battle going on between the two species in New Zealand. The South African import is becoming increasingly common in North Island gardens whereas the native mantid is less frequently seen. It is clear that the springbok species is far more aggressive; indeed, it has been seen to capture and kill members of the tangata whenua.

Noisy neighbours

THERE ARE A NUMBER of insect groups that communicate via acoustic signals to attract a mate or, if you want to put it in scientific lingo, their sounds are intentionally communicopulational. In the insect Order Orthoptera vocalisations appear to be particularly common. The word Orthoptera is derived from orthos, which means 'straight', and ptera, meaning 'wings'; in other words, insects with straight, hardened forewings and folded hindwings. It's a group that includes crickets, locusts, wetas, grasshoppers, mole crickets and katydids (pictured above).

Another characteristic of the Orthoptera is that they tend to possess large hind legs, adapted for jumping. These saltatorial habits are well known to those people who fear these insects unexpectedly jumping at them.

Black field cricket

THE BLACK FIELD CRICKET (*Teleogryllus commodus*; female pictured below) is noted for its wing-rubbing. This species is more vigorous in both its noise and actions than others of the Order Orthoptera. The male cricket is the singer. He slightly raises his hard, dark brown to black forewings, to create an amplifying sounding board, which helps to broadcast the noise he makes by rubbing the inner edges of his wings.

Adult females are, for a change, silent. They are recognised by their long, thin ovipositor, which sticks out at the back end.

NOISY NEIGHBOURS

The black field cricket occurs from Northland to about Marlborough. It may not be a major pest in the south, but up north it can be a real bother, particularly in dry summers and on heavier soils, prone to severe cracking. In some summers their damage to dairy pastures can run into the millions of dollars, simply by competing with the cattle for grass.

Black field crickets prefer a dark habitat during the day and so when the soil starts cracking, the crickets reside deep in the bowels of the earth, and only come out at night to feed. They avoid direct sunlight as their black bodies absorb heat from the sunshine, which can kill them very quickly. The cricket body starts to work like a miniature microwave oven, so the owners run the risk of getting cooked from the inside out!

Life cycle

The eggs hatch from around mid-October to the end of November. The little (4 millimetre long) larvae are jet black — just like the adults — but do not possess wings. They hide in and feed on the dense sward of grass and stay out of the sun. Kikuyu is a distinct favourite amongst the grass species. Other food items eagerly consumed are the flowers and seeds of a range of pasture plants.

The darlings grow rapidly, moulting their skin as they go. In order to recycle minerals and trace elements, the nymphs will eat their cast skins before getting back into the vegetable matter. 'Waste not, want not' is the motto here, and crickets often go one step further and cannibalise the bodies of their dead siblings, should the opportunity arise.

A cricket can moult as many as twelve times before it becomes a 30–35 millimetre long adult, with fully formed wings, in

DENSE POPULATIONS

A DECADE OR TWO AGO I WAS EMPLOYED BY MAF TO CARRY OUT CRICKET COUNTS AND DO OUTBREAK FORECASTS IN THE NORTHERN PART OF THE NORTH ISLAND. IT WAS NOT UNCOMMON TO MEASURE 50 OR 60 CRICKETS PER SQUARE METRE OF DAIRY PADDOCK — THE RECORD WAS 120! THIS GIVES SOME INDICATION AS TO THE POTENTIAL SEVERITY OF DAMAGE CAUSED BY THESE NOISY ORTHOPTERANS.

February. If you look at the half-grown nymphs (pictured opposite) in January, you can see their wing-buds appearing from their thoracic segments. With each moult, these wing-buds grow larger and larger.

Only the adults are able to mate and reproduce, and that's why they sing — or should I say 'chirp'? In any case, as soon as they are adults, black field crickets will be looking at securing the survival of their genetic material for the next generation. In autumn they lay their 2 millimetre long, light brown eggs (many hundreds per female) in cracks and holes in the ground, a perfect spot to spend the winter months in preparation for the hatch the following spring. That's when it all begins again — the 'one-generation' life cycle of the notorious pest of lawns and pasture in the warmer parts of New Zealand.

Impact

When crickets come out at night to feed they start off consuming and killing the grass plants adjacent to their home crack or hole, and slowly move further afield.

For lawn owners this creates sizeable bare patches of soil, ready to be invaded by weeds. For the dairy farmer in cricket-prone areas it means significantly less grass to eat for the primary milk producers, at a time when grass growth may have slowed down and herbage is in short supply. We are talking a measurable loss of butterfat production here!

Adult crickets eat a lot more grass 'per person' than small nymphs, which explains why cricket damage does not really show

up until mid to late summer. The adult can eat almost half its own body weight per night, and as these straight-winged singers can live for many weeks (right into May or June) their damage is often spectacular. Scientists have calculated that a very moderate infestation, averaging five crickets per square metre, is equivalent to having an extra stock unit in the paddock.

Apart from aesthetic and economic damage to the nation's expanse of lawn or dairy pasture, black field crickets can also make a nuisance of themselves indoors. They seldom fly far at night, but they can fly and will do so unexpectedly. Crickets have been found on ships up to 50 kilometres from the shore, no doubt attracted to bright lights onboard. Similarly, they can be lured into your lounge by means of lights shining through the ranchslider doors.

Once inside, they'll hide in cracks at the carpet's edge, and the males will keep singing for a mate, causing many a headache as they keep the baby awake. Their search for food leads them to vandalising soft furnishings and carpets — not a good look (despite the fact that you can be sure the offending insect will eventually die of dehydration).

Control

The remedy for gardeners is simple: grab yourself a full watering can and add a dash of dishwashing detergent. Pour this mixture into the cricket cracks in the middle of a hot summer's day, and watch for a few minutes. The crickets will race out of their hidey-holes (with vapour trails), rub the stinging soap out of their eyes while mumbling some abuse, and keel over on the bare soil next to the crack as the sun quickly cooks them.

Dairy farmers can use the same tactics to flush crickets out of selected cracks to get a feel for the average number of crickets per square metre. If numbers are higher than 10 individuals per square metre it is advisable to lay bait made from wheat, soaked in maldison: 250ml of Malathion 50% EC in 10kg wheat stirred together and soaked overnight in a cement mixer. Apply this bait at 10kg per hectare and watch the greedy gutses have a ball! A little grain can go a long way, as the poisoned insects will be consumed by cannibalistic siblings, who will then also become affected.

Dairy farmers should ideally be assessing their cricket populations in early to mid-January, before cricket damage becomes noticeably severe, and before the nymphs grow into egg-laying adults. That way they can prevent the heartache of seeing their grass disappear and also minimise egg numbers in the soil for the following year.

Watering cans and maldison-laced wheat-baits are not great options indoors, but any aerosol insecticide will flush out crickets in no time.

Katydid

FROM SUMMER TO WELL into late autumn, some Orthoptera singers attract attention in our gardens and lifestyle blocks. The most delicate of these is the katydid (*Caedicia simplex*, pictured below). All you need to do is go outside on a fine, still evening, just before sunset, and listen for that distinctive 'dzidits-tsits-zz' coming from the tops of shrubs and bushes. It's difficult to describe the sound, but someone suggested that the name katydid comes from the noise it makes: 'Katy did, she did . . .'

Maori named this insect kikihi ɔounamu, which roughly translates as 'the green one that makes a faint sound'. Indeed, the insect is bright green, both in the larval and adult stages. It walks sluggishly on the foliage and nibbles bits of vegetable matter here and there. Its parchment-like wings look a bit like leaves and allow it some form of sustained flight, but an F-16 it ain't. Katydids seem to love living in citrus bushes, plum trees and on rose plants, especially when the grasses and weeds underneath are long.

In the warmer parts of New Zealand, adult katydids will communicate well into winter, and can be found right into spring. Around the thermal areas of Rotorua it is not uncommon to see active adults on the tops of manuka bushes dziditsing away in the feeble midwinter afternoon sun.

Life cycle

The life cycle of this vegetarian is pretty straightforward, but a bit messy when it comes to timing. Eggs are laid in small rows along the stem of a suitable host plant. Each dark brown egg slightly overlaps the previous one. They are oval and measure about 3 millimetres long by perhaps 1.5 millimetres wide. The egg stage of a life cycle is often a very suitable vehicle for over-wintering, as it is resilient and can stand fluctuations in temperature and precipitation, and indeed, the katydid's eggs can be found during the winter months.

However adults, as well as young and older nymphs (larvae), are also able to

survive the colder months and can be found hiding on the food plants. The larvae are basically smaller versions of the green adults and often show developing wing-buds growing from the thoracic segments.

Impact

Economic plant damage is usually minimal — the insect tends to prefer fresh young growth and fruit. On roses, however, it can chew sizeable bites out of tender buds, which can be a bit of a heartbreak to dedicated rosarians. Interestingly, katydids often chew the inward-facing side of the buds and fruits, presumably because they like to stay under the cover of foliage as much as possible.

Control

These handsome native Orthopterans rarely need controlling, but if you find they are damaging your prize blooms, you'll know how and when to find them — go out with a torch in the evening and listen for their faint communications. Grab the offending katydid and, depending on your relationship with your neighbours, chuck it over the fence. Alternatively, you can set up a large breeding cage at home and study a colony of these individuals, because there are still some holes in our knowledge about these insects.

Weta

THE TERM WETA IS a Maori name, short for weta punga. They too belong to that fascinating insect order known as Orthoptera, and are therefore related to grasshoppers, crickets, locusts and katydids. I suppose that weta are simply long-horned grasshoppers, characterised by their long antennae and rather large hind legs.

Within the weta group we can distinguish two Families: the cave weta (Rhaphidophoridae) and the — for want of a better name — 'true' weta (Anosto-matidae). The true weta Family comprises more than 40 species, subdivided in four groups: tree weta, ground weta, tusked weta and the famous giant weta.

Life cycle

The long sword-like 'sting' is the female's ovipositor; her egg-laying tool! Many people think it's a vicious stinger, but in reality it is a very sensitive instrument, used to probe into soft soil with the specific purpose of laying eggs. A weta will never attack anyone with its ovipositor, although it may look sharp and mean.

The handy positioning of cerci, or feelers, at the tip of the abdomen becomes very apparent when you see weta mating (pictured below): the male has to be very careful in the vicinity of the ovipositor, and his cerci guide him safely to a fruitful reproductive session.

Observations of captive tree weta suggest that the females oviposit in autumn and the eggs hatch in spring. The small nymphs tend to disperse right away, which is not surprising as they can be subject to serious cannibalism by larger specimens, including their parents. It takes nine or 10 moults (and the best part of two years) for the nymphs to slowly reach adulthood.

During the later stages of development you can see the difference between males

and females: the latter form an ovipositor, which grows with each moult.

Often tree weta males develop an oversized head, with formidable mandibles. These weapons are very useful when picking fights with other territorial males, or when challenging for a hole in a tree, complete with harem. Rest assured that while you are asleep, fierce battles are raging in the foliage of your garden.

Impact

Weta are not known for causing significant damage to plants. Their foliage feeding results in typical chewing damage: holes and notches of irregular shape and size, the kind of stuff a plant can easily replace within a season.

The small weta that can sometimes be found in tunnels inside an apple are not necessarily the insects that caused the damage in the first place. More often than not a small weta nymph will take over the tunnel from a hatched codling moth.

Benefits

The most beneficial aspect of weta activity is the production of perfectly formed faecal pellets, which smell beautiful and act as slow-release fertilisers.

Parasites

Parasitic roundworms, protozoa and even mites can cause the weta a fair bit of discomfort, if not mortality. The most spectacular parasite is the gordian worm, a creature that develops inside the abdomen of a weta, and drives its dying host to a watery environment, where it completes its lifecycle. When the full-grown worm emerges from the weta's anal opening, it measures 30–40 millimetres long and about a millimetre wide. There are reports of four or five of such worms living inside just one weta!

How weta are infected by the larvae of this parasite is unclear; perhaps they need to feed on infected aquatic insects in order to take this parasite on board. In any case, it is not uncommon to find gordian worms in stagnant pools of water or inside weta.

Predators

No matter how large the jaws are or how hard a spiny back leg can kick, weta do have their enemies. Predation is a fact of life: tuatara, lizards, birds (morepork, kiwi, weka, kaka, saddleback etc) and bats are the original, native predators. Introduced rodents have taken this job into the new millennium.

Cave weta

CAVE WETA ARE THOSE large, long-legged critters you find sitting on the ceiling in dark tunnels and caves, but there are also a whole heap of much smaller species in hollow logs and under flakes of bark in forests. They are not commonly encountered by the average gardener or lifestyle-blocker, but are

NOISY NEIGHBOURS

nevertheless interesting invertebrates. At last count there were about 50 species of Rhaphidophoridae in New Zealand and the larger ones are known to be protein-feeders; even siblings are on the menu!

Tree weta and ground weta

IT IS YOUR LOCAL tree weta that inhabits the garden, makes a noise, and generally causes concern when it enters houses or unexpectedly emerges from a log of firewood. The ground weta is also pretty common in well-planted gardens, but the members of this group are usually small and non-threatening to us.

Your own backyard is a good starting point to find your local tree weta species. They occur through most of New Zealand, with the exception of the lower South Island and Stewart Island.

The best way to locate a tree weta (pictured below, in threat posture) is by going out at night and listening for their distinctive calls: a series of faint rasping noises at regular intervals. The sounds appear to come from the males, and are created by rubbing the inner thighs of the hind legs over a ribbed patch on the abdomen.

Tree weta have auditory pits or 'ears' situated on the 'forearms' of the pair of front legs (pictured opposite). If you want to get a good stereophonic impression of

NOT POISONOUS

ONE OF THE MOST PERSISTENT MYTHS ABOUT WETA IS THAT THE FEMALES ARE PARTICULARLY 'POISONOUS' AT CERTAIN TIMES OF THE YEAR, 'WHEN THEY DEFEND THEIR EGGS AND NEST'. IN FACT, WETA DO NOT HAVE ANY POISON, IN OR ON THEIR BODY, AT ALL. WHAT'S MORE, THE FEMALE HAS SUCH A SHORT MEMORY SPAN THAT SHE HASN'T GOT A CLUE WHERE THE EGGS WERE LAID AN HOUR AGO!

he was set to explain

he source of weta communication, it pays o have your ears as far apart as possible. Ears on the narrow head would not be as uitable as on the front legs, so it does make sense.

At the tip of the abdomen, right on he rear end of the tree weta, are some hort cerci. These appear to have the bility to pick up sound waves or vibrations; ears on your bum, in other vords. The cerci probably also register other worthwhile data for their owners, uch as meteorological information. Often ou'll see a weta sitting in the entrance to ts burrow, sampling the air with its cerci, before deciding whether to go out feeding or to have a quiet night inside.

From what I have observed they are omnivorous, preferring leaves of many different plant species, fruit, bark and even old proteinaceous rubbish (such as dead baby mice). They are also known to eadily eat insects, both dead and alive.

Tree weta have a sizeable crop, which cts as a storage organ and pre-digestion hamber for the food consumed. This means they do not have to go out to feed very night. Some specimens can stay in

bed for four or five nights in a row, after which they organise an enormous feeding frenzy for themselves — just to catch up.

Tree weta tend to be a gregarious lot. A male often gathers a small 'harem' of females and large juveniles, and maintains a loose bond with them; they can share burrows (or 'galleries') or live in separate tunnels. They often utilise hollow branches or stems, but bits of bamboo and the vacated tunnels of puriri moth caterpillars are also gratefully accepted as galleries.

Some folk see them hiding under corrugated iron, in letterboxes, or simply in a shrub with very dense foliage.

Tusked weta and giant weta

TUSKED AND GIANT WETA are as rare as hen's teeth, living mostly in inaccessible places such as mountaintops, cliffs, scree slopes, dense gorse bushes, and offshore islands without (rodent) predators. The giants are probably the

ones that terrify the ordinary, entomo-phobic punter, yet they should be the ones with the best public image. Nowhere else in the world has an Orthopteroid insect of such magnificent proportions (weighing as much as a song thrush), that occupies the ecological niche of a nibbling mammalian herbivore. The giant weta has been described as an 'invertebrate mouse' because of similarities in food ecology, but they are indeed gentle giants.

TAKE ON SOME TENANTS

PRUNING OR REMOVING HEDGES OR BUSHES FROM AROUND YOUR HOUSE CAN DEPRIVE WETA OF FOOD AND SHELTER. THEY WILL MOVE ON BUT UNTIL THEY DO, ANY WINDOW WITH A SMALL GAP AROUND THE FRAME IS A SUITABLE TEMPORARY GALLERY. IT WILL GIVE YOU A GOOD IDEA OF HOW MANY WETA HAVE BEEN LIVING IN THE IMMEDIATE VICINITY OF YOUR HOUSE. IF YOU WANT TO ENCOURAGE WETA IN YOUR GARDEN YOU CAN PROVIDE SHELTER BY HANGING ARTIFICIAL WETA-SHELTER SITES IN YOUR TREES. A PIECE OF BAMBOO, CUT OFF AT THE NODE ON ONE END, AND OPEN ON THE OTHER END, WILL DO NICELY AS A WETA HOUSE. JUST HANG IT SO THAT THE CLOSED END IS ON THE TOP OF THE TUBE. EXPERIMENT WITH VARIOUS WIDTHS OF BAMBOO, AND SOONER OR LATER YOU'LL GET SOME LODGERS

Stick insects

FRAGILE, DOCILE AND HARMLESS are not words you'd immediately associate with an invertebrate. Creepy-crawly insects have so often caused mayhem in our lives that their PR rating has gone downhill ever since we graduated from hunting and gathering to growing domesticated crops. Yep — we've had a few close shaves with these creatures and as a result we still regard insects as highly suspicious.

But some groups seem to have the natural ability to cause wonderment in humans. The fragile, docile and harmless stick insect is a very good example of such an insect.

There are about 2500 species of stick insects (Order Phasmatodea) in the world. In New Zealand we're looking at some 21 different species or subspecies and none of them is what you'd call spectacular.

In Aotearoa we do the basic types — long, thin, sometimes prickly, sometimes smooth, and in two colours, green and brown. But that doesn't mean they're boring or not worth looking at.

There are many stories about these insects — Maori had two names for them: ro and whe. Mind you, the same names were apparently used to indicate praying mantids as well, so there is some confusion as to who's who and what's what.

But one consistent story is that if a ro or a whe were to land on a pregnant woman, this occasion would be seen as a prediction as to which sex the child was going to be. I'm not aware of any experiments to test this hypothesis, so I can't comment on the phasmids' accuracy.

Another story is that stick insects can adapt themselves to their surroundings by changing colours. Hmmm. As an entomologist, what do you say when people are adamant that this is the truth?

Stick insects are all born green. Some species come in two colour 'morphs' — green and brown — and the brown phase tends to start kicking in at the second larval skin moult. Often only four or five per cent of a population attains a brown colour and once brown they stay brown forever. Green stick insects will stay green, whatever substrate they feed on. But they *do* change colour once in their lives and this change starts about two weeks before their death. Slowly these decrepit adults will turn a dull brown colour, before they let go of their perch and tumble off to the great twig in the sky.

Life cycle

Most stick insects hatch from eggs (pictured below) that have been lying on the soil during the winter. This can take place as early as late winter.

It is not easy, escaping from a small egg when you are endowed with long, gangly legs and stretched body parts. A lot of eggs do not even hatch and mortality is usually rather high, especially when conditions have been too dry.

Eggs are an important tool for identification. The taxonomy of stick insects has never been straightforward — males are especially difficult to classify — and eggs have recently been singled out as bearing a lot of useful characteristics that are specific to the various species.

They are, indeed, very ornate, with numerous 'ribs' and 'keels' and varying degrees of rugosity (wrinkles) and colour patterns on the outside. The capitulum is that funny little 'hat' on top of the egg; this pops open as the first instar larva emerges.

Once the larva is freed, it will move to a suitable food plant and start to nibble at the foliage. Stick insects are leaf chewers and they can do it with great gusto. Nocturnal or semi-nocturnal feeding behaviour appears to be the norm amongst phasmids.

There can be as few as four and as many as six or seven larval instars (pictured opposite) in as many months before adulthood is reached. That means a good number of skin moults and the same number of vulnerable periods.

Changing skin — like emerging from an egg — is a risky business. The new skin is super soft and if there's even *one* slight problem — such as a leg not wanting to come out of the old skin — the stick insect is doomed! The flip side of the coin carries some amazing biological capabilities for the stick insect. Should an immature specimen lose a leg — or two, or even three — to a predator, then these limbs can be partially regenerated during the following skin moults. The earlier in life regeneration commences, the bigger the new legs will grow.

The funny thing is that autotomy (literally: the shedding of a leg) is quite a natural process in phasmids. When they are

handled roughly or a predator takes a nip at one of the legs, the stick insect might first bleed or vomit a repugnant fluid to discourage the predator. But if that doesn't have the desired effect, autotomy is the next logical step in the escape process.

Phasmids generally rely on their cryptic colouration, which makes them look like something totally inedible. Indeed, they usually escape attention, especially when they sit very still.

Another trick they play is best described as 'playing possum'. They drop to the ground in a cataleptic state and just lie there, motionless. You can pick them up, give them a slight squeeze, or swing them around above your head — they simply won't move. Put them down and, after a while, they pick themselves up and start moving again, as if nothing's happened.

Food can be a bit of problem, perhaps not as much for a stick insect, but for researchers. You see, some species have really strong preferences for one particular food plant, say manuka. But breeding them on just that one host species alone often results in premature death and loss of your study objects. So obviously these creatures need a change of foliage from time to time.

If all goes well, the stick insect should mature by mid- to late summer, before the lethal frosts hit, and start looking for a partner to carry on the family name. But that in itself can be a bit of a problem.

A few species in New Zealand reproduce parthenogenetically. If a female can't find a male and lays unfertilised eggs, the offspring will all be female. If she lays fertilised eggs the resulting progeny will be a mixture of females and males.

STICK INSECTS

Impact

So, what's the problem?

The problem is in how the female lays her eggs. If she feeds on her favourite shrub with lots of lush new growth, she tends to drop her eggs straight down onto the ground. Dozens and dozens of eggs assemble at the base of the food source.

The following spring a lot of these eggs hatch and the offspring will feed on exactly the same shrub. But now the population is a lot bigger and the chewers can sometimes cause noticeable damage.

Even so, most folk don't really care about a partly denuded manuka, rimu, totara or rata. These insects are cute, curious and tangata whenua and they certainly give the impression of being fragile, docile and harmless, eh?

But . . . have you noticed by the way that most rose buds near these trees have been demolished by some vandal in the dead of night? Oh, you *had* noticed? Well . . . let me tell you a story about stick insects and alternative hosts . . . They are not *totally* docile and harmless.

Control

If you spot them, remove them from your prized plants and chuck 'em in a bucket. What you do with them after that depends entirely on the proximity of your local park or nature reserve, or your relationship with your neighbour!

WIDE VARIETY

THE LONGEST RECORDED STICK INSECT — SOME 50 CENTIMETRES LONG — WAS FOUND IN THE MALAYSIAN JUNGLE A DECADE AGO. THERE ARE ALSO SOME REAL THORNY DEVILS. THERE ARE SPECIES WITH SMALL, NON-FUNCTIONAL WINGS, SOME OF WHICH ARE REALLY COLOURFUL, AND THERE ARE THE FABULOUS LEAF INSECTS.

Rasping thrips

THRIPS ARE A REALLY common pest in New Zealand, especially in the warmer parts of the country and during the warmer parts of the year. The funny thing is, most people rarely see the actual insects — the damage is a lot easier to spot. Although it can be very distinctive, correct diagnosis of a thrips problem is so much easier when you know the background to their feeding behaviour.

Thysanoptera (meaning 'fringe-winged') are a strange group of critters. They are very small insects, with an elongated body and neatly folded wings held in a narrow strip over the abdomen. Looking at them under a microscope, you could be excused for likening them to miniature alligators. Adults are usually brown to dark brown or even black, whereas the nymphs tend to be a creamy-yellow in colour (pictured overleaf).

There are about 50 different species in New Zealand and not all of them are problematic — some feed on pollen and act as pollinators, whereas others are quite useful predators. Scientists have been having a close look at some of those beneficial species with a view to utilising their services in biological control. But the true rascals are the raspers.

Life cycle

Eggs are laid *somewhere* on the host plant (usually on the leaves, or in nooks and crannies of leaf buds). The larvae then rasp epidermal cells with gusto, day and night.

When you look with a hand lens at the nymphs of the greenhouse thrips you'll be able to see that they carry a crystal ball of dark brown liquid on the tip of their abdomen. This sticky poo is glued to the leaf surface from time to time, presumably when the external sewage becomes too heavy to carry.

Thysanoptera have a life cycle that is characterised by incomplete metamorphosis. This means that the development takes place without a pupa stage. The wings of the adult insect form gradually, with every larval skin moult. Often the last larval stages are non-feeding, resting stages that can be found on the host plant or in the soil, as in the case of onion thrips.

After about four larval feeding stages the dark-brown to black adults appear. These are just over a millimetre in length and show four light-coloured, thin wings, which are neatly folded over the abdominal segments. In mid-summer there will be a good number of them floating about. It always surprises me how these tiny, alligator-like insects can take to the sky with those flimsy and narrow wings, but given the right conditions they can and do.

Some species have a short (three- to four-week) life cycle and can produce many generations per year, especially in heated glasshouses.

Impact

Thrips, being very small insects, have short sucking snouts. There's no way that these mandibles will ever get anywhere near the sweet phloem juices that are available to the true suckers. That's why thrips have to make do with the fluids contained in individual cells. Each cell is rasped open and then sucked dry. That allows air to enter the damaged cells, giving the leaf surface a silvery or bronze sheen (pictured previous page), or a yellow stippled appearance. Greenhouse thrip damage is accompanied by copious amounts of dark brown sticky excrement and scratching marks. The destruction of epidermal cells can significantly reduce the photosynthetic surface area of leaves.

Control

Simple contact insecticides work well on most thrips that feed exposed on leaves, and especially on the nymphs. Diazinon, dichlorvos, Mavrik, maldison etc are all registered for use. The insecticide known as Spinosad (see page 14) has also shown good efficacy on thrips populations. On the nymphal stages, pyrethrum and mineral oil are two good organic alternatives, as is an abrasive fatty-acid spray.

Neem oil seems to work slowly by stopping the nymphs feeding. After a couple of applications a week apart, the nymphs will gradually fall off their leaves. The winged adults tend to react to the bad-tasting leaves much quicker and they fly off, cursing, to look for an uncontaminated leaf somewhere else.

The advantage of the slow-acting neem oil is that beneficial predators and parasites are not suddenly deprived of their hosts and are often able to complete their life cycle in relative comfort.

Western flower thrips are a lot more difficult to control with conventional insecticides because this species has shown to possess a good tolerance or resistance, especially to the regular systemics. If you are a commercial grower it may pay to contact your local adviser for details of chemical control or prevention — rotation of chemicals is recommended.

Most thrips over-winter in one form or another in crop debris, rubbish, or nooks and crannies in the soil, which means that hygiene is important to prevent successful hibernation. Weed control is also important, especially of the alternative host plants.

Western flower thrips have a knack of being able to use alternative hosts outside the glasshouse, so use very fine insect screens on glasshouse doors and windows and practise good hygiene: fumigate the soil with methyl bromide in between crops; obtain cuttings or plant material from a reputable supplier, free of this pest; do not plant flowers within 10 metres of the greenhouse; maintain a scorched-earth policy (concrete or asphalt surrounds); and regularly check nearby weed hosts. Yes, this pest is pretty bad!

In warm periods, adult thrips fly around looking for a suitable host to descend on. Their search is based on chemical cues (smell) and on colour patterns — perhaps the colour of flowers or leaves of a preferred host. It is whispered in the industry that thrips are very partial to the colour blue — a kind of mid sky-blue — and sticky traps designed for inside glasshouses make clever use of this colour. They not only catch large numbers of winged adults, but are also useful as monitoring tools for population build-up on the crop.

A word of warning, though: I'm not sure if it is such a good idea to use these blue traps in the open garden situation — blue is probably the best colour to attract small but useful insects as well as thrips. You see, when thrips are on the wing, so are their predators and tiny parasites. Timing is everything in nature, and when the prey is flying about, the control agents won't be far behind. Parasitic wasps and predatory hoverflies will no doubt come hovering by to inspect the giant blue, sticky 'flowers', and we don't want to immobilise the goodies, now do we?

Parasites

A parasitic wasp, *Thripobius semiluteus*, has recently been released in northern New Zealand to help control greenhouse thrips, especially on citrus.

Thrips predators come in a few different forms: *Amblyseius cucumeris* is a fast-moving brown predatory mite, specifically introduced for thrips control. Some ladybird beetle larvae will gobble up the odd thrips or two, and a large species of predatory thrips with black and white coloration, *Aeolothrips fasciatus*, has been observed eating its smaller cousins. A minute crabronid wasp catches thrips and stores them inside its nest inside the disused burrow of small woodborers. There the thrips bodies serve as food for the wasp's larvae.

Greenhouse thrips

THIS CREATURE WITH THE curious scientific name of *Heliothrips haemorrhoidalis* (a sun-loving thrip with piles?) has a habit of infesting citrus leaves and setting up shop on citrus fruit, especially where the fruit and a leaf touch each other. The resulting corky-grey surface will often split when the fruit expands.

The host range of the greenhouse thrips is huge and includes vegetables, fruit, and ornamentals. Common hosts are rhododendron, photinia, acmena, azalea, grape, citrus, kiwifruit and persimmon.

Western flower thrips

THE RESISTANT STRAIN OF this species (*Franklinierlla occidentalis*) has been established in New Zealand since 1992. Initially it was found in a few West Auckland glasshouses, but soon spread further afield and it will no doubt continue its march across the country.

It infests many different glasshouse crops (tomatoes, cucurbits, capsicum etc) and overseas there are 139 recorded host plants for this species: flowers, vegetables, ornamentals, white clover — you name it. Included in this list are the common weeds black nightshade (*Solanum nigrum*), cape gooseberry (*Physalis peruviana*) and Amaranthus. These weeds can act as important intermediate hosts, so take a good look at them from time to time and consider removing them as part of your control or prevention strategy.

Onion thrips

ONION THRIPS CAUSE LEAF-RASPING symptoms on a wide range of hosts: alliums, cucurbits, solanaceous crops, brassicas and beans. Strawberries are at risk too. A nasty habit of this species is the ability to vector spotted wilt virus through

tomato and capsicum crops, especially in glasshouses.

New Zealand flower thrips

THESE THRIPS DAMAGE FLOWERS by flecking the petals. They also go for a range of fruits, such as pipfruit, stone fruit and kiwifruit. Typical rasping damage results in scarring, russeting and discoloration, which make it a particularly problematic pest on stone fruit in Central Otago. To top it all off, it is endemic to New Zealand, and therefore a pest of quarantine importance on our export produce.

Gladiolus thrips

DAME EDNA'S BIGGEST FOE causes streaky, silvery damage on gladiolus leaves and blemishes on the flowers of this host. It's a clever pest that over-winters on the gladiolus corms.

Bottlebrush thrips

IN THE WARMER PARTS of New Zealand, where Australian bottlebrush is grown, young foliage can sometimes emerge totally twisted and distorted. This is the result of bottlebrush thrips damaging individual meristem cells in the growing tips of this plant. This host-specific pest hails originally from Australia.

Acknowledgements

THERE IS NO WAY anyone can write a book like this without help from family, friends and colleagues. The list of folk offering inspiration and information is too long to print.

I've communicated with experts from MAF, Hort Research and Landcare, colleagues at UNITEC, even respected entomologists and plant-health diagnosticians from as far away as the United States. They all gave me wonderful, raw information to work with. Any stuff-ups are entirely my own!

A lot of these stories are based on columns I wrote for the magazine *Growing Today*. Long-suffering editors Alan and Glenys have steered me in the right direction.

And talking about editors and 'steering': there's a whole nest of them at Random House (thank you Jenny, Fiona, Nic, Sarah and Katy). They had to cope with my crazy schedules.

Finally, my family needs a mention here. They put up with me tied to the computer and running around with camera gear at the oddest of times.

There's something quite magical when your 17-year-old son (Tristan) rings you just before midnight from a friend's place with the message: 'There is a cool native cockroach on the fence, shedding its skin; come over quickly with the camera!' It shows you that this book really is for everybody that shows an interest!

Index